The Illustrated Flora of Illinois

T0133264

The Illustrated Flora of Illinois
Robert H. Mohlenbrock, *General Editor*

The Illustrated Flora of Illinois

Flowering Plants

Asteraceae, Part 2

Robert H. Mohlenbrock

Illustrated by Paul W. Nelson

Southern
Illinois
University Press

Carbondale

Southern Illinois University Press

www.siupress.com

Copyright © 2019 by the Board of Trustees,
Southern Illinois University

Printed in the United States of America

22 21 20 19   4 3 2 1

Cover and title page illustration: *Helianthus grosseserratus* (sawtooth sunflower; cropped), by Paul
    W. Nelson

Library of Congress Cataloging-in-Publication Data
Names: Mohlenbrock, Robert H., 1931– author.
Title: Flowering plants. Asteraceae. Part 2 / Robert H. Mohlenbrock ; illustrated by Paul W.
    Nelson.
Other titles: Asteraceae | Illustrated flora of Illinois.
Description: Carbondale : Southern Illinois University Press, [2019] | Series: The illustrated flora
    of Illinois | Includes bibliographical references and index.
Identifiers: LCCN 2018038733 | ISBN 9780809337293 (pbk.) | ISBN 9780809337309 (e-book)
Subjects: LCSH: Compositae—Illinois. | Plants—Illinois.
Classification: LCC QK495.C74 M644 2019 | DDC 583/.983—dc23 LC record available at https://
    lccn.loc.gov/2018038733

Printed on recycled paper ♻

This book, my seventy-third, is dedicated to my dear wife, Beverly, and our three children—Mark, Wendy, and Trent—who have assisted me in so many various ways during my career.

Contents

Preface

Several volumes in the Illustrated Flora of Illinois series will be devoted to the dicotyledonous flowering plants; this volume is the ninth one. It follows publication of one on diatoms, one on ferns, six on monocotyledonous plants, and eight on other dicots.

The concept of the Illustrated Flora of Illinois is to produce a multivolumed flora of the plants of the state of Illinois. For each kind of plant known to occur in Illinois, complete descriptions and ecological notes will be provided, along with a statement of distribution.

There is no definite sequence for publication of the Illustrated Flora of Illinois. Volumes will appear as they are completed.

Herbaria from which specimens have been studied are located at Eastern Illinois University, the Field Museum of Natural History, the Gray Herbarium of Harvard University, the Illinois Natural History Survey, the Illinois State Museum, the Missouri Botanical Garden, the Morton Arboretum, the New York Botanical Garden, the Shawnee National Forest, Southern Illinois University Carbondale, the United States National Herbarium, the University of Illinois, and Western Illinois University. In addition, some private collections have been examined. The author is indebted to the curators and staff of these herbaria for the courtesies extended.

I am deeply grateful to Henry and Alice Barkhausen and the Gaylord and Dorothy Donnelley Foundation for their generous support that made this volume possible. Paul W. Nelson meticulously prepared all of the illustrations. Hybrids are not illustrated. My wife, Beverly, assisted me in several of the herbaria and typed drafts of the manuscript. Madison Preece Olsen prepared the indices and the glossary. Without the help of all these individuals and organizations, this book would not have been possible.

The Illustrated Flora of Illinois

Flowering Plants:
Asteraceae, Part 2

County Map of Illinois

Introduction

Flowering plants that form two "seed leaves," or cotyledons, when the seed germinates, are called dicotyledons, or dicots. These far exceed the number of species of monocots, or flowering plants that produce a single "seed leaf" upon germination. This is the ninth volume of the Illustrated Flora of Illinois to be devoted to the dicots of Illinois.

The system of classification adopted for the Illustrated Flora of Illinois was proposed by Thorne in 1968 and modified in 1999 and 2007. This system is a marked departure from the more familiar system of Engler and Prantl. This latter system, which is still followed in many regional floras, is out of date and does not reflect the vast information recently gained from the study of cytology, biochemistry, anatomy, and embryology. In fact, the Thorne system no longer depicts many of the relationships adhered to in 1968.

Since the arrangement of orders and families proposed by Thorne is unfamiliar to many, an outline of the orders and families of flowering plants known to occur in Illinois is presented.

Those names in boldface are orders and families that have already been published in this series. The family in capital letters is the one described in this volume of the Illustrated Flora of Illinois.

Order Annonales
Family Magnoliaceae
Family Annonaceae
Family Calycanthaceae
Family Aristolochiaceae
Family Lauraceae
Family Saururaceae

Order Berberidales
Family Menispermaceae
Family Ranunculaceae
Family Berberidaceae
Family Papaveraceae

Order Nymphaeales
Family Nymphaeaceae
Family Ceratophyllaceae

Order Sarraceniales
Family Sarraceniaceae

Order Theales
Family Aquifoliaceae
Family Hypericaceae

Family Elatinaceae
Family Ericaceae

Order Ebenales
Family Ebenaceae
Family Styracaceae
Family Sapotaceae

Order Primulales
Family Primulaceae

Order Cistales
Family Violaceae
Family Cistaceae
Family Passifloraceae
Family Cucurbitaceae
Family Loasaceae

Order Salicales
Family Salicaceae

Order Tamaricales
Family Tamaricaeae

Order Capparidales
Family Capparidaceae
Family Resedaceae
Family Brassicaceae

Order Malvales
Family Sterculiaceae
Family Tiliaceae
Family Malvaceae

Order Urticales
Family Ulmaceae
Family Moraceae
Family Urticaceae

Order Rhamnales
Family Rhamnaceae
Family Elaeagnaceae

Order Euphorbiales
Family Thymelaeaceae
Family Euphorbiaceae

Order Solanales
Family Solanaceae
Family Convolvulaceae
Family Cuscutaceae
Family Polemoniaceae

Order Campanulales
Family Campanulaceae

Order Santalales
Family Celastraceae
Family Santalaceae
Family Loranthaceae

Order Oleales
Family Oleaceae

Order Geraniales
Family Linaceae
Family Zygophyllaceae
Family Oxalidaceae
Family Geraniaceae
Family Balsaminaceae
Family Limnanthaceae
Family Polygalaceae

Order Rutales
Family Rutaceae
Family Simaroubaceae
Family Anacardiaceae
Family Sapindaceae
Family Aceraceae
Family Hippocastanaceae
Family Juglandaceae

Order Myricales
Family Myricaceae

Order Chenopodiales
Family Phytolaccaceae
Family Nyctaginaceae
Family Aizoaceae
Family Cactaceae
Family Portulacaceae
Family Chenopodiaceae
Family Amaranthaceae
Family Caryophyllaceae
Family Polygonaceae

Order Hamamelidales
Family Hamamelidaceae
Family Platanaceae

Order Fagales
Family Fagaceae
Family Betulaceae
Family Corylaceae

Order Rosales
Family Rosaceae
Family Mimosaceae
Family Caesalpiniaceae
Family Fabaceae
Family Crassulaceae
Family Penthoraceae
Family Saxifragaceae
Family Droseraceae
Family Staphyleaceae

Order Myrtales
Family Lythraceae
Family Melastomaceae
Family Onagraceae

Order Gentianales
Family Loganiaceae
Family Rubiaceae
Family Apocynaceae
Family Asclepiadaceae
Family Gentianaceae
Family Menyanthaceae

Order Bignoniales
Family Bignoniaceae
Family Martyniaceae
Family Scrophulariaceae
Family Paulowniaceae
Family Plantaginaceae
Family Orobanchaceae
Family Lentibulariaceae
Family Acanthaceae

Order Cornales
Family Vitaceae
Family Nyssaceae

Family Cornaceae
Family Haloragidaceae
Family Hippuridaceae
Family Araliaceae
Family Apiaceae

Order Dipsacales
Family Caprifoliaceae
Family Adoxaceae
Family Valerianaceae
Family Dipsacaceae

Order Lamiales
Faimily Hydrophyllaceae
Family Heliotropaceae
Family Boraginaceae
Family Verbenaceae
Family Phrymaceae
Family Callitrichaceae
Family Lamiaceae

ORDER ASTERALES
FAMILY ASTERACEAE

Since only one part of one family is included in this book, no general key to the dicot families has been provided. The reader is invited to use my companion book, *Guide to the Vascular Flora of Illinois* (2013) for keys to all families of flowering plants in Illinois.

The nomenclature for the species and lesser taxa used in this volume has been arrived at after lengthy study of recent floras and monographs. Synonyms, with complete author citations, that have been applied to species and lesser taxa in Illinois are given under each species. A description, while not necessarily intended to be complete, covers the important features of the species.

The common name, or names, is the one used locally in Illinois. The habitat designation is not always the habitat throughout the range of the species, but only for it in Illinois. The overall range for each species is given from the northeastern to the northwestern extremities, south to the southwestern limit, then eastward to the southeastern limit. The range has been compiled from various sources, including examination of herbarium material and field studies of my own. A general statement is given concerning the range of each species in Illinois.

The distribution has been compiled from extensive field study as well as herbarium study. The author is indebted to the curators and staffs of the herbaria for the courtesies extended.

The second volume of the Asteraceae of Illinois consists only of the tribes Heliantheae and Senecioneae. The characteristics of the Heliantheae are mixed. Plants may have opposite, alternate, or basal leaves; radiate or discoid heads; phyllaries in 2–5 series; receptacles that are flat, convex, or columnar; and either paleate or epaleate ray flowers; usually only pistillate but sometimes neutral disc flowers; usually perfect and fertile cypselae either flat or pyramidal or angled; and pappus, if any, of bristles, scales, or short awns. The Senecioneae are characterized by discoid or radiate heads, bisexual disc flowers, alternate leaves, phyllaries in 1–2 series, and pappus of 30 or more capillary bristles.

Descriptions and Illustrations

This is the second of three volumes describing the family Asteraceae in Illinois.

This family for many years was called the Compositae. Some botanists in the past, with sound reasoning, have divided the family into three families, but I have chosen to follow the current belief that these three segregates should be treated as a single but diverse family.

In Illinois, I am recognizing 391 species in 120 genera, as well as 27 hybrids and 73 lesser taxa, making it the largest family of flowering plants in the state. Of these 391 species, 131 of them are non-natives, or roughly 33 percent. The Poaceae, or grass family, the second largest in Illinois, has 357 species in 103 genera, along with 5 hybrids and 44 lesser taxa.

Worldwide, members of the Asteraceae exhibit nearly every growth form, but in Illinois, all species are herbaceous and include annuals, biennials, and perennials. Only one species, *Mikania scandens*, is a vine. All the rest are either erect, ascending, or prostrate herbs. Some species have latex. Many plants have taproots and/or fibrous roots, although there are a number of rhizomatous or stoloniferous species.

With such a large and diverse family, the species exhibit every type of leaf possible. A few species have only basal leaves; others have only cauline leaves; still others have both basal and cauline leaves. In that latter group, the basal leaves may or may not be present at flowering time. In those plants with cauline leaves, the leaves may be alternate, opposite, or whorled. Leaves may be simple, unlobed or lobed, or variously compound. Nearly every type of pubescence may be found in the family.

Flowers occur in heads, known as capitulae, with each head having few to numerous flowers. The heads are usually several and are arranged in a variety of inflorescences, called arrays, that may be in the form of corymbs, cymes, panicles, thyrses, or racemes. Occasionally the head may be reduced to one per plant. Each head is surrounded by one or more series of bracts, known as phyllaries. The phyllaries make up the involucre. In a given head, the phyllaries may be equal or unequal. They may be green and herbaceous or they may have a scarious margin. They may be glabrous or variously pubescent and glandular or eglandular. In *Coreopsis* and *Taraxacum*, there are tiny bractlets, called calyculi, at the outside base of the phyllaries.

Heads may have ray flowers or disc flowers or both. Ray flowers have rays that are zygomorphic and are often erose or shallowly lobed at the tip. They may have 5 stamens and an inferior ovary. Rays are sometimes referred to as petals by the uninformed. Disc flowers are actinomorphic, usually short-tubular, with 4 or usually 5 lobes. They may have 5 stamens and an inferior ovary.

Experts in this family recognize different types of heads. Radiate heads have peripheral rays that may be pistillate or sterile and central flowers that are disc flowers that are either bisexual or staminate. Liguliferous heads have only ray flowers that are bisexual. Discoid heads have only disc flowers that may be bisexual, only

7

pistillate, or only staminate. In disciform heads, all flowers are disc flowers, but the peripheral flowers have filiform corollas that are usually only pistillate.

All flowers in a head share a common receptacle. The receptacle may be flat or convex. It may bear tiny scales called paleae. The paleae of a receptacle are referred to as chaff. Receptacles without paleae are said to be epaleate or naked. When the paleae are shed, they may leave either a smooth or pitted receptacle. Occasional hairs or bristles may also be present on the receptacle.

Although the fruits of the Asteraceae are often referred to as achenes, they are actually cypselae. Achenes are dry, one-seeded fruits that are derived from flowers with a unicarpellate superior ovary. Cypselae are dry, one-seeded fruits that are derived from flowers with a bicarpellate inferior ovary. The cypselae may be crowned or subtended by pappus, which may be in the form of capillary or plumose bristles, awns, or scales. The pappus is usually thought to be the remains of a calyx.

Although molecular phylogenetic studies of the Asteraceae have resulted in different classifications within the family, I am following the more traditional divisions of the family into tribes used in *Flora of North America*.

Below are listed the tribes of Asteraceae in North America, along with the number of species in North America and the number of species in Illinois.

Tribe Mutisieae	7 genera, 14 species in U.S.—None in Illinois
Tribe Cynareae	17 genera, 116 species in U.S.
	10 genera, 34 species, 1 hybrid in Illinois
Tribe Arctotideae	3 genera, 4 species in U.S.—None in Illinois
Tribe Vernonieae	6 genera, 25 species in U.S.
	2 genera, 6 species, 1 hybrid in Illinois
Tribe Cichorieae	50 genera, 230 species in U.S.
	21 genera, 51 species, 5 hybrids in Illinois
Tribe Calenduleae	4 genera, 7 species in U.S.—None in Illinois
Tribe Gnaphalieae	19 genera, 101 species in U.S.
	5 genera, 10 species in Illinois
Tribe Inuleae	3 genera, 5 species in U.S.
	1 genus, 1 species in Illinois
Tribe Senecioneae	29 genera, 167 species in U.S.
	8 genera, 17 species in Illinois
Tribe Plucheae	3 genera, 12 species in U.S.
	1 genus, 2 species in Illinois
Tribe Anthemideae	26 genera, 99 species in U.S.
	9 genera, 24 species in Illinois
Tribe Astereae	77 genera, 719 species in U.S.
	19 genera, 103 species, 2 hybrids in Illinois
Tribe Heliantheae	150 genera, 746 species in U.S.
	42 genera, 141 species, 20 hybrids in Illinois
Tribe Eupatorieae	170 genera, 250 species in U.S.
	8 genera, 25 species, and 5 hybrids in Illinois

Because of the large number of taxa of Asteraceae in Illinois, the treatment of them will be in three volumes. Since goldenrods and asters are the largest groups and are confusing, I have elected to treat them in the initial volume, along with other members of the tribe Astereae. However, in order to have each of the three volumes more or less comparable in size, I added the tribe Anthemideae to the first volume.

I do not want to divide any tribe into more than one volume, so the tribes Heliantheae and Senecioneae comprise this second volume, although it will be treating a few more taxa than the other volumes.

The third book in this series encompasses the other tribes in the Asteraceae: Cynareae, Vernonieae, Gnaphalieae, Inuleae, Eupatorieae, Plucheae, and Cichorieae.

After the description of the Asteraceae and the key to all the genera of Asteraceae in Illinois, the description, habitat notes, nomenclatural issues, uses, and other applicable information will be provided for each taxon, along with illustrations. However, hybrids are not illustrated. Following the name of each taxon are any synonyms that may be pertinent.

Key to the Genera of Asteraceae in Illinois

Names appearing in boldface are treated in this book; those plants treated in the first volume (*Asteraceae, Part 1*) are in italics followed by I; those plants treated in the third volume (*Asteraceae, Part 3*) are in italics followed by III.

1. Flowering heads with only ray flowers; latex present . . . . .Group 1
1. Flowering heads with disc flowers, the ray flowers present or absent; latex absent.
  2. All the leaves, or at least those on the lower part of the stem, opposite or whorled
  . . . . Group 2
  2. None of the leaves opposite or whorled.
    3. Ray flowers present.
      4. Rays yellow or orange. . . . . .Group 3
      4. Rays blue, purple, pink, rose, or white, not yellow or orange . . . . Group 4
    3. Ray flowers absent.
      5. Leaves simple, entire, toothed, or shallowly lobed . . . . .Group 5
      5. Leaves deeply pinnatifid or pinnately compound . . . . Group 6

Group 1

Flowering heads with only ray flowers; latex present.
1. Flowering heads blue, purple, or pinkish.
  2. Cypselae beakless.
    3. Pappus a crown of scales in 2 to several series.
      4. Pappus of blunt scales; cauline leaves present . . . . 100. *Cichorium* III
      4. Pappus of aristate scales; cauline leaves absent . . . . 101. *Catananche* III
    3. Pappus of capillary bristles in 1 series . . . . 109. *Nabalus* III
  2. Cypselae with a beak about 0.5 mm long up to a long filiform beak.
    5. Pappus of capillary bristles; heads several; peduncles subtended by bractlets.
      6. Beak of cypselae about 0.5 mm long; phyllaries in 1–2 series . . . .
      . . . . 107. *Mulgedium* III

  6. Beak of cypselae 1–6 mm long; phyllaries in 1 series. . . . . 108. *Lactuca* III

  5. Pappus of plumose bristles; head solitary; peduncles not subtended by bractlets. . .

. . . . 116. *Tragopogon* III

1. Flowering heads yellow, orange, cream, or whitish.

 7. Flowering heads cream or whitish . . . . 109. *Nabalus* III

 7. Flowering heads yellow or orange.

  8. Cypselae with a beak 0.5–12.0 mm long.

   9. Cypselae with a stout beak about 0.5 mm long.

    10. Pappus of capillary bristles; heads usually numerous; bractlets at base of

     peduncle up to 10 . . . . 108. *Lactuca* III

    10. Pappus of plumose bristles; heads 1 to very few; bractlets at base of

     peduncle up to 20 . . . . 112. *Leontodon* III

   9. Cypselae with a filiform beak 1–12 mm long.

    11. Receptacle paleate; pappus of 2 kinds, the outer of capillary bristles, the

     inner of plumose bristles . . . . .113. *Hypochaeris* III

    11. Receptacle epaleate; pappus not as above.

     12. Outer pappus of scales, inner pappus of capillary bristles; flowering

      heads 1 to few . . . . 120. *Pyrrhopappus* III

     12. Pappus of all uniform bristles; heads solitary or numerous.

      13. Pappus of capillary bristles.

       14. Stems scapose . . . . .103. *Taraxacum* III

       14. Stems with some cauline leaves . . . . 104. *Chondrilla* III

      13. Pappus of plumose bristles.

       15. Flowering head solitary, 4–8 cm across. . . . . . . 116. *Tragopogon* III

       15. Flowering heads numerous, up to 2 cm across.

        16. Peduncles subtended by 5 foliaceous bractlets. . . . .

. . . . 114. *Helminotheca* III

        16. Peduncles subtended by numerous narrow bractlets. . . . .

. . . . 115. *Picris* III

  8. Cypselae beakless.

   17. Leaves with spinescent teeth; pappus in 4 series . . . . 110. *Sonchus* III

   17. Leaves not spinescent; pappus in 1–2 series, or absent.

    18. Pappus absent.

     19. Flowering heads usually solitary; peduncles not subtended by bractlets

. . . . 119. *Serinia* III

     19. Flowering heads few to several; peduncles subtended by 4–5 bractlets .

. . . . 105. *Lapsana* III

    18. Pappus present.

     20. Outer pappus of scales, inner pappus of bristles.

      21. Phyllaries up to 35, in 2–5 series; cypselae with 10 ribs . . . .

. . . . 117. *Nothocalais* III

      21. Phyllaries up to 18, in 1–2 series; cypselae with 10–20 ribs . . . .

. . . . .118. *Krigia* III

     20. All pappus of bristles.

      22. Pappus of plumose bristles. . . . . 112. *Leontodon* III

      22. Pappus of capillary bristles.

       23. None of the leaves pinnatifid. . . . . 111. *Hieracium* III

       23. Some of the leaves pinnatifid.

24. Peduncles subtended by 5–12 bractlets; phyllaries up to 16 per head; involucre 4–15 mm across . . . . .102. *Crepis* III
24. Peduncles subtended by 3–5 bractlets; phyllaries 8 per head; involucre 2–3 mm across . . . . 106. *Youngia* III

Group 2

Flowering heads with disc flowers, the ray flowers present or absent; all the leaves, or at least those on the lower part of the stem, opposite or whorled; latex absent.
1. Ray flowers present.
  2. Rays yellow or orange.
    3. Pappus of numerous capillary bristles; plants creeping . . 51. *Calyptocarpus*, p. 131
    3. Pappus of awns, scales, 1–3 small bristles, or absent; plants upright (procumbent in *Sanvitalia*).
      4. Leaves simple, entire, serrate, shallowly lobed, or palmately lobed.
        5. Leaves shallowly palmately lobed; phyllaries in 1 series . . . . .35. *Smallanthus*, p. 46
        5. Leaves not palmately lobed; phyllaries in 2 or more series.
          6. Disc flowers sterile with poorly developed cypselae . . . .43. *Silphium*, p. 89
          6. Disc flowers fertile with well-developed cypselae.
            7. Ray flowers persistent on the cypselae . . . . 41. *Heliopsis*, p. 84
            7. Ray flowers deciduous from the cypselae.
              8. Pappus absent or of 1–3 small bristles.
                9. Petioles absent or nearly so; cypselae 4–5 mm long, strongly angular; pappus completely absent . . . . 37. *Guizotia*, p. 50
                9. Petioles at least 3 mm long; cypselae 1.0–2.5 mm long, weakly angular; pappus absent or of 1–3 small bristles . . . . .50. *Acmella*, p. 129
              8. Pappus of awns or scales, rarely absent.
                10. Cypselae flat.
                  11. Stems winged.
                    12. Cypselae wingless; all leaves opposite . . . . 45. *Verbesina*, p. 116
                    12. Cypselae winged; some of the leaves alternate . . . . .46. *Actinomeris*, p. 119
                  11. Stems wingless . . . . 47. *Ximenesia*, p. 123
                10. Cypselae angular.
                  13. Cypselae 3-angled; pappus of cypselae of ray flowers with 3 awns . . . . .42. *Sanvitalia*, p. 88
                  13. Cypselae 2- or 4-angled; pappus of cypselae of ray flowers with 2 or 4 awns.
                    14. Cypselae with 2 or 4 stout, barbed awns . . . . 57. *Bidens*, p. 200
                    14. Cypselae with 2 small, barbless awns, or pappus rarely absent.
                      15. Rays usually 8 per head; phyllaries in 2 series . . . . 54. *Coreopsis*, p. 177
                      15. Rays of various numbers per head, not all of them 8; phyllaries in several series.

36. Pappus absent; receptacle paleate; stems square . . . . 48. *Melanthera*, p. 125
36. Pappus of capillary bristles; receptacle epaleate; stems not square.
    37. Some or all the leaves whorled; flowers purple or rose . . . .
    . . . . 73. *Eutrochium* III
    37. Leaves opposite (if rarely whorled, the flowers white); flower white, blue, or pink.
        38. Receptacle conical; flowers blue . . . . 74. *Conoclinium* III
        38. Receptacle flat; flowers white or pink.
            39. Flowers pink . . . . 78. *Fleischmannia* III
            39. Flowers white.
                40. Phyllaries all of same length . . . . 80. *Ageratina* III
                40. Phyllaries of different lengths . . . . 72. *Eupatorium* III

Group 3

Plants with both ray and disc flowers; rays yellow or orange; leaves alternate or basal; latex absent.

1. Leaves simple, entire, serrate, or shallowly lobed.
    2. Most or all the leaves basal.
        3. Head solitary.
            4. Pappus of 5–8 translucent scales . . . . 64. **Tetraneuris**, p. 248
            4. Pappus of capillary bristles . . . . 70. **Tussilago**, p. 283
        3. Heads several . . . . 66. **Packera**, p. 256
    2. Most of the leaves cauline.
        5. Rays reflexed; disc long-columnar or globose.
            6. Disc long-columnar; cauline leaves clasping . . . . 38. **Dracopis**, p. 52
            6. Disc globose; cauline leaves not clasping . . . . 63. **Helenium**, p. 240
        5. Rays spreading, rarely slightly reflexed; disc globose, conical, or flat.
            7. Pappus entirely of capillary or barbellate bristles.
                8. Leaves spinulose-dentate . . . . .17. *Prionopsis* I
                8. Leaves not spinulose-dentate.
                    9. Upper leaves clasping . . . . 86. *Inula* III
                    9. None of the leaves clasping.
                        10. Inflorescence more or less a flat-topped corymb.
                            11. Leaves up to 5(–12) mm wide, some of them usually glandular-punctate . . . . 7. *Euthamia* I
                            11. Leaves usually more than 5 mm wide, not glandular-punctate . . .
                            . . . . 9. *Oligoneuron* I
                      10. Inflorescence paniculate, thyrsoid, or in axillary clusters . . . .
                      . . . . 8. *Solidago* I
        7. Pappus of a short crown, awns, or scales or, if capillary bristles present, scales are present in addition.
            12. Receptacle bristly; pappus of 6–10 awned scales . . . . .62. **Gaillardia**, p. 235
            12. Receptacle epaleate or paleate; pappus not as above.
                13. Pappus of the cypselae of the disc flowers scaly, with short teeth, a crown, or absent.
                  14. Outer pappus scaly, inner pappus bristly on the cypselae of the disc flowers.
                      15. Cypselae of ray flowers thick, of disc flowers flat. . .11. *Heterotheca* I
                      15. All cypselae flat . . . . 10. *Chrysopsis* I

14. Pappus of the cypselae of the disc flowers with short teeth, a crown, or absent.
    16. Disc flowers sterile; phyllaries in 2–3 series . . .43. *Silphium*, p. 89
    16. Disc flowers fertile; phyllaries in several series. . . . .
    . . . . .39. *Rudbeckia*, p. 54
13. Pappus of the cypselae of the disc flowers with 2–8 awns.
    17. Phyllaries gummy. . . . . .18. *Grindelia* I
    17. Phyllaries not gummy.
        18. Pappus of 2 awns, these sometimes deciduous.
            19. Disc flowers sterile; phyllaries in 2–3 series . . . .
            . . . . .43. *Silphium*, p. 89
            19. Disc flowers perfect; phyllaries in several series.
                20. Cypselae wingless . . . . 52. *Helianthus*, p. 133
                20. Cypselae winged.
                    21. Stems unwinged . . . . 47. *Ximenesia*, p. 123
                    21. Stems winged. . . . . .46. *Actinomeris*, p. 119
        18. Pappus of 5 or more awns, persistent.
            22. Disc flowers sterile . . . . .5. *Amphiachyris* I
            22. Disc flowers fertile . . . . 6. *Gutierrezia* I
1. Leaves simple and deeply divided, or leaves compound.
  23. Receptacle epaleate.
    24. Rays reflexed; pappus reduced to a short crown, or absent; phyllaries in 2
    series. . . . . 40. *Ratibida*, p. 80
    24. Rays spreading; pappus of barbellate bristles; phyllaries in 1 series.
        25. Stems glandular-pubescent; rays 1–3 mm long . . . . 65. *Senecio*, p. 251
        25. Stems eglandular; rays 6–10 mm long.
            26. Leaves once-pinnatifid. . . . . .66. *Packera*, p. 256
            26. Leaves 2- to 3-pinnatifid . . . . 65. *Senecio*, p. 251
  23. Receptacle paleate.
    27. Receptacle flat; phyllaries in 2–3 series; disc flowers sterile . . 43. *Silphium*, p. 89
    27. Receptacle conical; phyllaries in several series; disc flowers fertile.
        28. Leaves 2- to 3-pinnate or–pinnatifid . . . . 27. *Cota* I
        28. Leaves lobed or 1-pinnatifid . . . . .39. *Rudbeckia*, p. 54

Group 4
Plants with both ray flowers and disc flowers; rays not orange nor yellow; latex absent; leaves alternate or basal.
1. Leaves simple and entire, serrate, or shallowly lobed (sometimes with a basal pair of pinnae in *Tanacetum*).
  2. Rays blue, purple, or pink.
    3. Rays reflexed; disc conical . . . . .44. *Echinacea*, p. 108
    3. Rays spreading; disc flat or subglobose.
        4. Pappus a double series of capillary bristles . . . . .4. *Ionactis* I
        4. Pappus a single series of capillary bristles, awns, scales, or absent.
            5. Leaves mostly basal; flower heads 1 to few . . . . .2. *Bellis* I
            5. Leaves mostly cauline; flower heads usually numerous.
                6. Pappus of capillary bristles.
                    7. Rays usually more than 50 . . . . 12. *Erigeron* I
                    7. Rays usually less than 50.

8. Basal leaves cordate; inflorescence a flat-topped corymb . . . .
. . . . 15. *Eurybia* I
8. Basal leaves tapering or rounded at the base or, if cordate, the inflorecence racemose or paniculate.
  9. Basal leaves cordate . . . . 19. *Symphyotrichum* I
  9. Basal leaves tapering or rounded at the base, or absent at flowering time.
    10. Annuals. . . . . 19. *Symphyotrichum* I
    10. Perennials.
      11. Leaves sericeous on both surfaces . . . .
      . . . . 19. *Symphyotrichum* I
      11. Leaves glabrous or pubescent, but not sericeous.
        12. Stems glandular, at least above; leaves oblong; phyllaries densely glandular. . . . .
        . . . . 19. *Symphyotrichum* I
        12. Stems eglandular; leaves linear or lanceolate or elliptic; phyllaries eglandular.
          13. Leaves coarsely toothed, all of them 2.5 cm wide or wider. . . . . 1. *Aster* I
          13. Leaves entire or sparsely and finely toothed, many of them less than 2.5 cm wide . . . .
          . . . . 19. *Symphyotrichum* I
6. Pappus of short awns or scales.
  14. Pappus of 6–10 awned scales; rays purple; phyllaries in 2–3 series; receptacle setose . . . . 62. **Gaillardia**, p. 235
  14. Pappus of 2 awns and tiny bristles; rays pink; phyllaries in several series; receptacle epaleate . . . . 14. *Boltonia* I
2. Rays white.
  15. Pappus of capillary bristles.
    16. Pappus in a double series.
      17. Inflorescence flat-topped . . . . 3. *Doellingeria* I
      17. Inflorescence paniculate . . . . 4. *Ionactis* I
    16. Pappus in a single series.
      18. Rays less than 5 mm long. . . . . 13. *Conyza* I
      18. Rays more than 5 mm long.
        19. Inflorescence flat-topped . . . . 9. *Oligoneuron* I
        19. Inflorescence paniculate, not flat-topped.
          20. Rays usually more than 50 . . . . 12. *Erigeron* I
          20. Rays usually fewer than 50. . . . . 19. *Symphyotrichum* I
15. Pappus of 2–3 awns, a short crown, or absent.
  21. Pappus absent; cypselae 3-angled. 23. *Chamaemelum* I
  21. Pappus present; cypselae not 3-angled.
    22. Pappus reduced to a short crown; leaves sometimes with a basal pair of pinnae . . . . 21. *Tanacetum* I
    22. Pappus of 2–3 awns; leaves without a basal pair of pinnae.
      23. Stems winged . . . . 45. **Verbesina**, p. 116
      23. Stems unwinged.
        24. Pappus of 2 awns and tiny bristles; plants glabrous; receptacle naked . . . . 14. *Boltonia* I

24. Pappus of 2–3 awns, without bristles; plants pubescent;
    receptacle paleate . . . . 32. *Parthenium*, p. 37
1. Leaves deeply pinnatifid or 1- to 3-pinnate.
  25. Leaves merely deeply lobed or pinnatisect.
    26. Pappus absent; receptacle naked. .29. *Leucanthemum* I
    26. Pappus or 2–3 awns; receptacle paleate. . . . . 32. *Parthenium*, p. 37
  25. Leaves 1- to 3-pinnate.
    27. Pappus of capillary bristles; phyllaries more or less squarrose; flowers blue or
    purple . . . . .16. *Machaeranthera* I
    27. Pappus a low crown, or absent; phyllaries not squarrose; flowers white.
      28. Receptacle paleate, phyllaries in several series; pappus absent.
        29. Plants aromatic; receptacle conical . . . . 25. *Anthemis* I
        29. Plants not aromatic; receptacle flat .22. *Achillea* I
      28. Receptacle naked; phyllaries in 2–3 series; pappus a low crown, or absent.
        30. Cypselae 3-ribbed; plants not aromatic . . . . 28. *Tripleurospermum* I
        30. Cypselae 5-ribbed; plants aromatic . . . . 26. *Matricaria* I

Group 5

Flowering heads with only disc flowers; leaves alternate or basal and simple,
entire, serrate, shallowly lobed, or pinnatifid (or with a basal pair of pinnae in
Tanacetum); latex absent.
1. Leaves with spine-tipped teeth.
  2. Flowers yellow . . . . .97.*Centaurea* III
  2. Flowers purple, pink, pale blue, or white.
    3. Pappus of barbellate bristles; phyllaries in 2 series; stems winged. . . . .
    . . . . 89. *Onopordum* III
    3. Pappus of simple bristles, plumose bristles, or a crown of scales; phyllaries in
    several series; stems unwinged.
      4. Pappus a crown of scales; flowers pale blue . . . . 88. *Echinops* III
      4. Pappus of bristles; flowers purple, white, or pink.
        5. Pappus of plumose bristles. . . . . 91. *Cirsium* III
        5. Pappus of simple bristles . . . . 90. *Carduus* III
1. Leaves without spine-tipped teeth.
  6. Outer row of disc flowers appearing ligulate.
    7. Outer phyllaries entire, never spine-tipped.
      8. Phyllaries coriaceous, yellowish, without a hyaline margin . . . 94. *Amberboa* III
      8. Phyllaries thin, green, with a hyaline margin . . . . 93. *Acroptilon* III
    7. Outer phyllaries fimbriate or laciniate, some of them spine-tipped.
      9. Involucre 3–4 cm high; pappus 6–12 mm long . . . . .95. *Plectocephalus* III
      9. Involucre 1.0–1.5 mm high; pappus 3 mm long or less . . . . 97. *Centaurea* III
  6. Outer row of disc flowers tubular, not appearing ligulate.
    10. Flowers greenish.
      11. Phyllaries and fruits with hooked bristles, prickly or appearing prickly.
        12. Flowers unisexual . . . . 31. **Xanthium**, p. 29
        12. Flowers perfect. . . . . 92. *Arctium* III
      11. Phyllaries and fruits without hooked bristles, not prickly . . . . . 24. *Artemisia* I
    10. Flowers white, purple, cream, yellow, orange, blue, pink, brownish, or rusty.
      13. Phyllaries and fruits with hooked bristles . . . . 92. *Arctium* III
      13. Phyllaries and fruits without hooked bristles.

14. Flowers orange; pappus absent or of short scales; some leaves clasping; receptacle paleate . . . . .96. *Carthamus* III
14. Flowers not orange; pappus of capillary bristles, a short crown, or 5–8 scales; leaves not clasping; receptacle epaleate.
   15. Flowers yellow, cream, brownish, or rusty.
      16. Phyllaries in 1 series; flowers yellow . . . . 65. **Senecio,** p. 251
      16. Phyllaries in several series; flowers cream, brownish, or rusty.
         17. Pappus of plumose bristles . . . . 76. *Brickellia* III
         17. Pappus of capillary bristles.
            18. Heads leafy bracted . . . . .84. *Gnaphalium* III
            18. Heads not leafy bracted . . . . 82. *Pseudognaphalium* III
   15. Flowers pink, purple, or white.
      19. Flowers pink or purple.
         20. Flowers in glomerules, subtended by a 3-lobed bract . . . .
            . . . . .98. *Elephantopus* III
         20. Flowers and bracts not as above.
            21. Phyllaries in 1 series, subtended by calyx-like bracts; leaves basal . . . . 71. **Petasites,** p. 285
            21. Phyllaries in several series; leaves not basal.
               22. Pappus of plumose or barbellate bristles . . .77. *Liatris* III
               22. Pappus of simple bristles.
                  23. Pappus in a double series. . . . . 99. *Vernonia* III
                  23. Pappus in a single series.
                     24. Capillary bristles united at base . . . .
                        . . . . .85. *Gamochaeta* III
                     24. Capillary bristles not united at base . . . .
                        . . . . 87. *Pluchea* III
      19. Flowers white.
         25. Phyllaries in 1 series.
            26. Calyx-like bracts at base of phyllaries; leaves basal . . . .
               . . . . 71. **Petasites,** p. 285
            26. Calyx-like bracts absent; leaves cauline.
               27. Phyllaries 5; flowers 5 per head. .69. **Arnoglossum,** p. 279
               27. Phyllaries more than 5; flowers more than 5 per head . . . .
                  . . . . .67. **Erechtites,** p. 272
         25. Phyllaries in 2-several series.
            28. Some of the leaves hastate . . . . 68. **Hasteola,** p. 277
            28. None of the leaves hastate.
               29. Pappus a short crown . . . . 21. *Tanacetum* I
               29. Pappus of capillary bristles.
                  30. Phyllaries in 2 series, not scarious. . . 20. *Brachyactis* I
                  30. Phyllaries in several series, scarious.
                     31. Most of the leaves near the base of the plant . . . . . .
                        . . . . .81. *Antennaria* III
                     31. Most of the leaves cauline. . . . . . . .83. *Anaphalis* III

Group 6
Flowering heads with only disc flowers; leaves alternate or basal, pinnatifid to pinnately compound; latex absent.

1. Flowers greenish.
  2. Pistillate involucre with 2 flowers and 2 beaks; receptacle paleate . . . .
  . . . . 30. *Ambrosia*, p. 110
  2. Involucres not as above; receptacle naked or woolly . . . . 24. *Artemisia* I
1. Flowers yellow or greenish yellow.
  3. Pappus of capillary bristles; phyllaries in 1 series . . . . 65. *Senecio*, p. 251
  3. Pappus of 12–20 hyaline scales, of short awns, or absent; phyllaries in 2–4 series.
    4. Pappus of 12–20 hyaline scales. . . . . 61. *Hymenopappus*, p. 233
    4. Pappus a short crown, or absent.
      5. Receptacle conic; plants aromatic. . . . . 26. *Matricaria* I
      5. Receptacle flat; plants not aromatic . . . . 21. *Tanacetum* I

Tribe Heliantheae Cass.

Annuals, biennials, or perennials (in Illinois); leaves usually cauline, mostly opposite but occasionally alternate or whorled, entire to dentate to pinnatifid to compound; heads usually with ray flowers and disc flowers, less commonly with disc
flowers only, arranged in various types of inflorescences; phyllaries in 2–5 series,
unequal or subequal; receptacle flat to convex to conic to columnar, paleate or
epaleate, sometimes pitted; ray flowers usually pistillate, sometimes neutral; disc
flowers usually bisexual, fertile, the corolla usually 5-lobed; cypselae flat, pyramidal, prismatic, or angled; pappus of capillary bristles, scales, short awns, or absent.
  Worldwide, this tribe consists of about 300 genera and more than 3,300 species.

Key to the Subtribes of Heliantheae in Illinois
1. Receptacles epaleate (small setae may be present in Gaillardiinae.)
  2. Ray flowers absent; cypselae 4-angled and with 12–16 ribs . . . .
  . . . . Subtribe Hymenopappiinae
  2. Ray flowers present; cypselae not 4-angled and without 12–16 ribs . . . .
  . . . . Subtribe Gaillardiinae
1. Receptacles paleate.
  3. Heads discoid; anthers distinct . . . . Subtribe Ambrosiinae
  3. Heads or most of them radiate; anthers connate.
    4. Bractlets present on the peduncles; cypselae flat or 4-angled. . . . .
    . . . . Subtribe Coreopsidiinae
    4. Bractlets absent; cypselae not flat nor 4-angled (3- or 4-angled in Milleriinae.)
      5. Phyllaries caducous.
        6. Disc flowers perfect and fertile.
          7. Pappus of scales or bristles . . . . .Subtribe Galinsogiinae
          7. Pappus absent . . . . Subtribe Milleriinae
        6. Disc flowers staminate . . . . Subtribe Ambrosiinae
      5. Phyllaries persistent.
        8. Receptacle high-conic to columnar.
          9. Outer phyllaries longer than the inner ones; ray flowers neutral . . . .
          . . . . .Subtribe Rudbeckiinae
          9. Outer phyllaries shorter than the inner ones; ray flowers perfect, fertile. . .
          . . . . Subtribe Ecliptiinae
        8. Receptacle flat to convex to low-conic.

10. Ray flowers neutral. . . . . Subtribe Helianthinae
10. Ray flowers pistillate, fertile.
    11. Disc flowers staminate.
       12. Ray flowers 7–13; disc flowers 40–80; cypselae with 30–40 . . . . . .
       . . . . Subtribe Melampodiinae
       12. Ray flowers 2–6; disc flowers 12–30; cypselae with 3–6 ribs . . . . . .
       . . . . .Subtribe Polymniinae
    11. Disc flowers perfect, fertile. . . . . Subtribe Ecliptiinae

Subtribe Ambrosiinae Less.

Annual or perennial herbs (in Illinois); leaves cauline (in Illinois), alternate or opposite, sessile, sometimes lobed, usually pubescent; heads discoid or radiate, arranged in various kinds of inflorescences; involucres of various shapes, sometimes forming spiny burs; phyllaries few to numerous, in 1–8 series, usually unequal; receptacles usually flat to convex, usually with paleae; ray flowers 0 or 5, pistillate, fertile; disc flowers 5–60 or more, staminate, the corolla 5-lobed; anthers free from each other or connate; cypselae various; pappus absent.

    Worldwide there are 12 genera and about 75 species in subtribe Ambrosiinae, found in the subtropical or warm temperate regions of the New World. In Illinois, there are five genera, eighteen species, two hybrids, and two lesser taxa.

Key to the Genera of Tribe Heliantheae Subtribe Ambrosiinae in Illinois

1. Heads radiate . . . . .32. *Parthenium*
1. Heads discoid.
    2. Most of the leaves opposite; fruit not a large prickly bur.
       3. Pistllate involucre nut-like or a small bur; leaves cleft or with a pair of teeth near the base . . . . 30. *Ambrosia*
       3. Pistillate involucre neither nut-like nor a small bur; leaves undivided.
          4. Heads in racemose spikes, bracteate. . . . . .33. *Iva*
          4. Heads in paniculate spikes, ebracteate. . . . . 34. *Cyclachaena*
    2. Most of the leaves alternate; fruit a large prickly bur. . . . . 31. *Xanthium*

30. Ambrosia L.—Ragweed

Annual or perennial herbs; stems erect (in Illinois), branched; leaves cauline, opposite or alternate, simple, often deeply lobed; heads discoid, unisexual, the staminate in terminal spikes or racemes, the pistillate in the axils of the uppermost leaves; involucre of staminate flowers cup-shaped to saucer-shaped, to 5 mm across, the phyllaries up to 15, connate in a single series; involucre of pistillate flowers with 10–80 phyllaries in one series, the outer free or connate, the inner connate, forming tubercles or spines; receptacle flat or convex, usually paleate; corolla white or purplish, funnelform, 5-lobed; cypselae ovoid, oblongoid, or fusiform, glabrous, tuberculate, spiny, or bur-like; pappus absent.

    Ambrosia consists of about 40 species, mostly in temperate and tropical regions of the New World. *Franseria* and *Gaertneria* are genera that at one time were considered distinct from *Ambrosia* but are now considered by most to belong to *Ambrosia*.

This genus is distinguished by its green, unisexual heads, with the staminate flowers borne in terminal spikes or racemes, and the pistillate spikes borne in the axils of the uppermost leaves.

1. All leaves alternate, bipinnatifid ....1. *A. tomentosa*
1. At least the lowest leaves, and sometimes all the leaves, opposite.
  2. Leaves not deeply divided, usually with a pair of teeth near the base...2. *A. bidentata*
  2. Most leaves deeply divided.
    3. Leaves either palmately 3- or 5-lobed or entire, but never pinnatifid... 3. *A. trifida*
    3. Some or all the leaves pinnatifid or bipinnatifid.
      4. All leaves pinnatifid or bipinnatifid.
        5. Leaves on petioles up to 3 cm long; burs with 5–7 pointed spines up to 0.8 mm long; annual with fibrous roots................... 4. *A. artemisiifolia*
        5. Leaves sessile or subsessile, harshly pubescent above; involucre unarmed or with blunt tubercles or with pointed spines about 0.2–0.6 mm long; perennial from creeping rootstocks.
          6. Involucre with blunt tubercles; stems harshly scabrous .. 5. *A. psilostachya*
          6. Involucre with pointed spines; stems somewhat scabrous ....
          ....6. *A. X intergradiens*
      4. Some leaves palmately 3- or 5-lobed, other pinnatifid or bipinnatifid ....
      .... 7. *A. X helenae*

1. **Ambrosia tomentosa** Nutt. Gen. N. Am. Pl. 1:186. 1818. Fig. 1.
Franseria discolor Nutt. Trans. Am. Phil. Soc. II, 7:345. 1841.
Gaertneria discolor (Nutt.) Kuntze, Rev. Gen. Pl. 339. 1891.
Franseria tomentosa (Nutt.) A. Nels. New Man. Bot. Cent. Rocky Mt. 542. 1909.

Perennial herb from deep roots; stems ascending to erect, to 40 cm tall, tomentose; leaves cauline, alternate, bipinnatifid, acute at the apex, tapering to the petiolate base, to 10 cm long, to 5 cm wide, the lobes dentate, glabrous or nearly so on the upper surface, gray-tomentose on the lower surface, the petiole obscurely winged; staminate heads in racemes on pubescent peduncles up to 1 cm long, with 25 or more flowers; involucre of staminate heads saucer-shaped, to 5 mm across, the phyllaries 10–15, connate, strigillose; pistillate heads few in the axils of the upper leaves, with numerous unequal phyllaries; burs pyriform, flattened, 1.5–3.5 mm long, scabrous to tomentose, with 8–10 short, conic spines to 1 mm long; pappus absent.

Common Name: False ragweed.
Habitat: Disturbed soil.
Range: Native to the western United States; mostly naturalized east of the Mississippi River.
Illinois Distribution: Known only from LaSalle and McHenry counties.

The common name of false ragweed is derived from the fact that this species was at one time placed in genera other than *Ambrosia*, the ragweeds.

This is the only species of *Ambrosia* in Illinois that has leaves that are all alternate.

Ambrosia tomentosa flowers from July to October.

1. *Ambrosia tomentosa* (False ragweed).

a. Habit.
b. Staminate head.
c. Pistillate involucre.

d. Disc flower.
e. Fruiting involucre.

2. Ambrosia bidentata Michx. Fl. Bor. Am. 2:182. 1803. Fig. 2.

Annual herbs from fibrous roots; stems erect, branched, to 1 m tall, hirsute; leaves mostly alternate, lanceolate, acuminate at the apex, rounded to cordate at the sessile base, up to 3.5 cm long, up to 8 mm wide, entire or serrate and usually with two large teeth at the base, hispid on both surfaces, glandular; staminate heads sessile in dense spikes, with 6–8 flowers; involucre of staminate heads

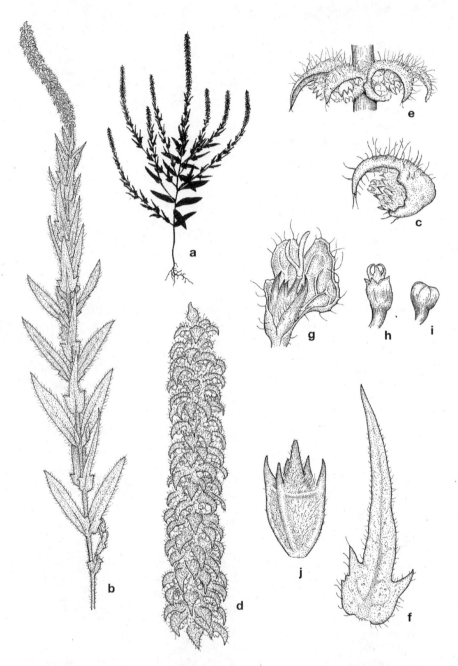

2. *Ambrosia bidentata*
   (Toothed ragweed).
a. Habit.
b. Upper part of
   habit.

c. Staminate involucre.
d. Staminate raceme.
e. Cluster of staminate
   involucres.
f. Leaf.

g, h. Pistillate involucres.
i. Unopened pistillate
   involucre.
j. Fruiting involucre.

cup-shaped but with a reflexed lobe, 2.5–4.0 mm across, hispid; pistillate heads few in the axils of the upper leaves, with numerous phyllaries; burs ovoid to ellipsoid, 4-angled, 5–8 mm long, hispid, with 4 spines up to 1 mm long; pappus absent.

Common Name: Toothed ragweed; lance-leaved ragweed.
Habitat: Fields, prairies, disturbed sites.
Range: Ohio to Minnesota, south to Texas and Georgia; also Connecticut.
Illinois Distribution: Occasional in the southern three-fifths of Illinois.

This ragweed is easily distinguished by its leaves with a pair of hastate teeth near the base.

An unnamed hybrid, presumably between *A. bidentata* and *A. trifida*, was described by Gray in 1886. The type was collected by H. Eggert in St. Clair County. It has also been found in Clay, Jersey, Sangamon, and Vermilion counties.

Flowering time for *Ambrosia bidentata* is July to October.

3. **Ambrosia trifida** L. Sp. Pl. 2:987. 1753.
Robust annual herbs from fibrous roots; stems stout, branched, to 6 m tall, usually hispid, rarely glabrous or nearly so; leaves mostly opposite, or the uppermost sometimes alternate, mostly ovate, acute to acuminate at the apex, rounded or truncate at the petiolate base, to 20 cm long, to 20 cm wide, 3-nerved, usually 3-cleft, occasionally 5-cleft, or less commonly uncleft, scabrous, glandular, sometimes decurrent on the petioles; staminate heads in terminal racemes, the pedicels to 3 mm long, with up to 25 flowers; involucre of staminate heads saucer-shaped, 2–4 mm in diameter, scabrous; pistillate heads few in the axils of the uppermost leaves, with numerous phyllaries; burs obovoid, 3–12 mm long, usually glabrous, with 4–5 pointed spines or sometimes with blunt tubercles; pappus absent.

Two varieties occur in Illinois:
a. Petioles of at least the upper leaves wing-margined; burs 6–12 mm long, with short,
   pointed spines . . . . 3a. *A. trifida* var. *trifida*
a. Petioles wingless; burs 3–7 mm long, with blunt tubercles . . . . 3b. *A. trifida* var. *texana*

3a. **Ambrosia trifida** L. var. **trifida** Figs. 3 (a–c), 130 (a) (see p. 288).
Ambrosia integrifolia Muhl. ex Willd. Sp. Pl. 4:375. 1805.
Ambrosia trifida L. var. *integrifolia* (Muhl. ex Willd.) Torr. & Gray, Fl. N. Am. 2:290.
   1841.
Ambrosia trifida L. f. *integrifolia* (Muhl. ex Willd.) Fern. Rhodora 40:347. 1938.

Petioles of at least the upper leaves wing-margined; burs 6–12 mm long, with short, pointed spines.

Common Name: Giant ragweed.
Habitat: Floodplain forests, disturbed areas.
Range: In every state, except Hawaii, and in most Canadian provinces.
Illinois Distribution: Common throughout the state; in every county.

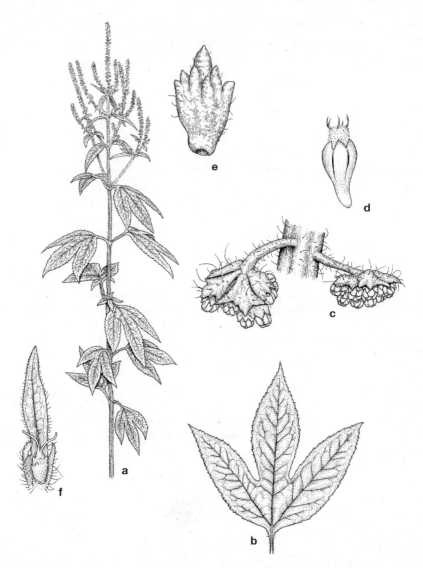

3. *Ambrosia trifida* var. *trifida* (Giant ragweed).
a. Upper part of plant.
b. Leaf.
c. Clusters of staminate flowers. var. *texana* (Texas ragweed).
d. Pistillate flower.
e. Fruiting involucre.
f. Pistillate flower with phyllary.

Some plants without lobed leaves occur. They have been called var. *integrifolia*. The leaves of this variety have a strong resemblance to the leaves of *Iva annua*, but the *Iva* has linear bracts subtending each staminate head.

Ambrosia trifida was originally a plant of floodplain forests where it sometimes occurred in large stands. However, when towns and cities became established, this species moved into disturbed habitats of these municipalities and became rank weeds. Although its original habitat was in the floodplains, it is now considered to be a noxious weed in Illinois where it is a major cause of autumnal hay fever. South of Murphysboro, along the Big Muddy River, there remains about a five-acre stand of *Ambrosia trifida*, which is how I envision it must have looked to the first surveyors who were in the area in the early 1800s. Specimens with unlobed leaves may be called f. *integrifolia*.

Plants as tall as six meters have been observed in Illinois.

Ambrosia trifida flowers from July to October.

3b. **Ambrosia trifida** L. var. **texana** Scheele, Linnaea 22:156. 1849. Fig. 3 (d, e). Petioles wingless; burs 3–7 mm long, with blunt tubercles.

Common Name: Texas ragweed.
Habitat: Disturbed soil (in Illinois).
Range: Native to the southwestern United States; sometimes adventive elsewhere.
Illinois Distribution: Occasional in the southern fourth of Illinois.

The wingless petioles and the blunt spines on the burs are distinctive. When I was teaching a wetland plant class in Texas a few years ago and our van was heading down a highway, I pointed out a large stand of giant ragweed we observed out the window. One of the students said he called the plant bloodweed. Since I had never heard this, I asked him why he called it bloodweed. We stopped the van and walked over to the ragweed patch. He made a vertical scratch on the stem of the plant with his thumbnail. A red-purple color immediately surfaced in the scratch. Through the years, I have done this procedure in several parts of Texas, and often the red-purple color appears. In the few plants of this variety I have seen in southern Illinois, the red-purple color usually appears, while all of our plants of the typical variety lack this red-purple color.

This variety flowers from July to October.

4. **Ambrosia artemisiifolia** L. Sp. Pl. 2:988. 1753. Fig. 4.
Ambrosia elatior L. Sp. Pl. 2:987–988. 1753.
Ambrosia artemisiifolia L. var. *elatior* (L.) Desc. Fl. Med. Antilles 1:239. 1821.
Ambrosia artemisiifolia L. f. villosa Fern. & Grisc. Rhodora 37:135. 1935.

Annual herbs from fibrous roots; stems erect, branched, to 2 m tall, pubescence sparse and spreading or nearly glabrous; leaves mostly opposite but often with some of the uppermost leaves alternate, pinnatifid to bipinnatifid to tripinnatifid, or the uppermost linear and entire, tapering to the petiolate base, glabrous or sparsely pubescent, not scabrous, to 5 cm long, to 4 cm wide, glandular, the petioles 0.3–3.0 cm long; staminate heads in terminal racemes, with 12–20 flowers; involucre of staminate heads cup-shaped, 2–3 cm across, glabrous or pubescent; pistillate heads 2–15 in the axils of the uppermost leaves; burs obovoid, 4–5 mm long, sparsely pubescent, with 5–7 short pointed spines up to 0.8 mm long.

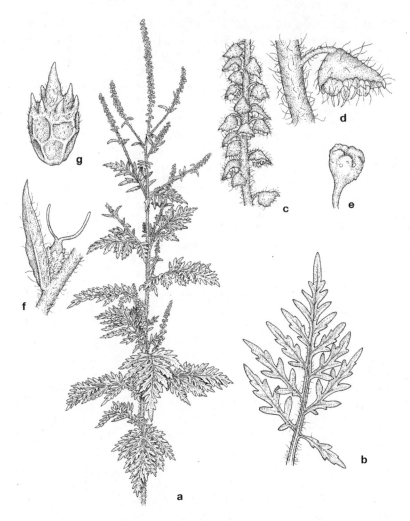

4. *Ambrosia artemisiifolia* (Common ragweed).
a. Habit.
b. Leaf.

c. Staminate raceme.
d. Staminate involucre.
e. Unopened pistillate involucre.

f. Pistillate involucre with subtending bract.
g. Fruiting involucre.

Common Name: Common ragweed.
Habitat: Fields, disturbed soils.
Range: Throughout the United States and Canada; in every state except Hawaii.
Illinois Distribution: Common throughout the state; in every county.

Our plants actually belong to var. *elatior*. Typical var. *artemisiifolia* usually has some leaves unlobed and no leaves tripinnatifid. Plants with pilose leaves and stems may be called f. *villosa*.

This is a common plant in open habitats. It flowers from August to October.

5. **Ambrosia X intergradiens** W. H. Wagner, Rhodora 60:198. 1958. Not illustrated.

Perennial herbs from long-creeping rootstocks; stems erect, branched, to 1.5 cm tall, somewhat scabrous, with densely spreading hairs; leaves opposite, or the uppermost alternate, pinnatifid or bipinnatifid, to 6 cm long, to 3.5 cm wide, tapering to the subsessile or petiolate base, scabrous, glandular, the petioles 0.5–1.7 cm long; staminate heads in terminal racemes, with 5 to many flowers; involucre of staminate heads cup-shaped, 2.0–4.5 mm across, hispid; pistillate heads 1–8 in the axils of the uppermost leaves; burs ovoid or obovoid, 3–6 mm long, pubescent with pointed spines 0.2–0.6 mm long; pappus absent.

Common Name: Hybrid ragweed.
Habitat: Disturbed, often sandy soil.
Range: Illinois, Michigan, Minnesota, Wisconsin.
Illinois Distribution: Scattered in the northern half of the state.

This hybrid appears to be more closely related to *Ambrosia psilostachya*, particularly in the fact that both are perennial with creeping rootstocks. It is presumed to be the hybrid between *A. artemisiifolia* and *A. psilostachya*.

This hybrid flowers from July to October.

6. **Ambrosia psilostachya** DC. Prodr. 5:526. 1836. Fig. 5.
Ambrosia coronopifolia Torr. & Gray, Fl. N. Am. 2:291. 1842.

Perennial herbs from long-creeping rootstocks; stems erect, branched, to 1.5 m tall, harshly scabrous, with densely appressed hairs; leaves opposite, or the uppermost alternate, pinnatifid or bipinnatifid, to 6 cm long, to 3.5 cm wide, tapering to the sessile or subsessile base, scabrous, hispid, glandular, the petioles 0.5–1.4 cm long; staminate heads in terminal racemes, with 5-many flowers; involucre of staminate heads cup-shaped, 2.0–4.5 mm across, usually hispid; pistillate heads 1–3 in the axils of the uppermost leaves; burs ovoid or obovoid, 3–6 mm long, pubescent with blunt tubercles up to 0.5 mm long, or tubercles absent; pappus absent.

Common Name: Western ragweed.
Habitat: Sandy soil in fields and disturbed areas.
Range: In every state and all of the southern Canadian provinces.
Illinois Distribution: Occasional in the northern three-fourths of Illinois; also St. Clair and Williamson counties.

This species looks very similar to *Ambrosia artemisiifolia* but is a perennial with long-creeping rhizomes and with burs with blunt tubercles or no tubercles at all.

7. **Ambrosia X helenae** Rowlee, Nat. Can. 71:272. 1944. Not illustrated.
Annual herb from fibrous roots; stems erect, branched, to 2.5 m tall, usually somewhat scabrous; leaves opposite, or the uppermost sometimes alternate, some

of them palmately 3- or 5-lobed, some of them pinnatifid to bipinnatifid, to 8 cm long, to 6 cm wide, tapering to the petiolate base, usually scabrous on the upper surface, usually pubescent on the lower surface, glandular; staminate heads in terminal racemes, with 15–20 flowers; involucre of staminate heads cup-shaped, 2–3 mm across, usually sparsely pubescent; pistillate heads few, in the axils of the uppermost leaves; burs obovoid, 6–10 mm long, glabrous or nearly so, with 5–6 pointed or blunt tubercles; pappus absent.

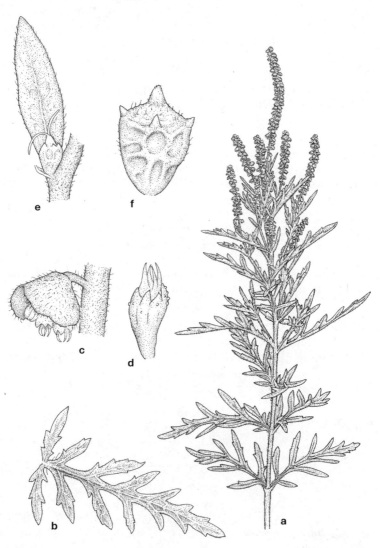

5. *Ambrosia psilostachya* (Western ragweed).
a. Habit.

b. Leaf.
c. Staminate involucre.
d. Unopened pistillate involucre.

e. Pistillate involucre with subtending bract.
f. Fruiting involucre.

Common Name: Helen's ragweed.
Habitat: Disturbed soil.
Range: Scattered throughout the United States.
Illinois Distribution: Known only from Champaign County.

This is the hybrid between *A. trifida* and *A. artemisiifolia*, and the leaves are a good indication of this. Some of these are palmately 3- or 5-lobed, resembling the leaves of *A. trifida*, and others are pinnatifid or bipinnatifid, resembling the leaves of *A. artemisiifolia*.

This hybrid flowers from July to October.

31. **Xanthium** L—Cocklebur

Annuals from fibrous roots; stems erect, branched, pubescent; leaves alternate except for sometimes the very lowermost opposite, petiolate; heads discoid, the staminate singly or in spikes or racemes, the pistillate below; staminate heads with involucres saucer-shaped, with 6–15 or more phyllaries in 1–2 series; pistillate head with involucres ellipsoid, with 30 or more phyllaries in 6–12 series, the outermost free, the innermost connate, straight or usually uncinate at the tip, at maturity becoming a hard prickly bur; receptacle conic to columnar, paleate; pistillate heads with 20–50 flowers, the corolla whitish, funnelform, 5-lobed; cypselae fusiform, enclosed by the prickly bur, the bur with 1–2 beaks at the tip; pappus absent.

I recognize about twelve species, most native to the New World but introduced worldwide, one in Europe, and one in tropical America.

Since members of *Xanthium* are unattractive, with obnoxious prickly burs, few botanists have actually studied them in much detail. The most definitive treatment, and the one I follow to some degree, was prepared by Millspaugh and Sherff in 1919. Their delineation of species is based primarily on the intricate characteristics of the mature fruiting bur.

More recently, most botanists have discounted these fruiting intricacies and recognize only two species. I believe, however, that these fruiting characteristics are important; therefore, I am admitting eight species of *Xanthium* in the Illinois flora. Most of these may be found in the southern fourth of the state, although a few of them occur statewide.

All botanists agree that *X. spinosum* is distinct because of its tapering leaf base and the presence of stiff axillary spines. Plants with cordate or truncate leaf bases and no axillary spines are usually called *X. strumarium*. In analyzing the mature fruits, however, I divide these latter plants into seven species in Illinois. In fact, typical *X. strumosum*, which I believe is native to Europe, is not known from Illinois.

Xanthium strumarium has pilose burs only 8–14 mm long and 5–7 mm thick, prickles about 15–50 in one view at a time, beaks of the burs only 1–2 mm long, and straight prickles only 2–3 mm long. I have seen no specimens from Illinois with these characteristics.

At maturity, the burs cling readily to clothing or to the fur of mammals, providing an effective method of seed dispersal.

The chart compares and contrasts the species of the *X. strumarium* complex in Illinois.

Key to the Species of *Xanthium* in Illinois

1. Leaves lanceolate, tapering to the base, with 3-parted axillary spines. . . . .
. . . . 1. *X. spinosum*
1. Leaves ovate, cordate or truncate at the base, without axillary spines.
    2. Burs 30–40 mm long, with more than 200 prickles in view at one time . . . .
    . . . . 2. *X. speciosum*
    2. Burs up to 30 mm long, with up to 200 prickles in view at one time.
        3. Burs and prickles glabrous or puberulent, but not hispid, pilose, or villous-hirsute; beaks of fruits straight or nearly so.
            4. Burs 20–25 mm long, oblongoid to ellipsoid, more than twice as long as thick. . . . . 3. *X. chinense*
            4. Burs up to 20 mm long, ovoid, up to twice as long as thick.
                5. Leaves more or less thin; burs 10–18 mm thick; prickles 50–60 in view at one time, 4–6 mm long. . . . . 4. *X. globosum*
                5. Leaves thick; burs 6–9 mm thick; prickles 100–200 in view at one time, 2.0–3.5 mm long . . . . 5. *X. chasei*
        3. Burs and prickles hispid, pilose, or villous-hirsute; beaks of burs usually uncinate.
            6. Burs hispid, with 50–80 prickles in view at one time; leaves thin . . . .
            . . . . 6. *X. inflexum*
            6. Burs pilose or villous-hirsute, with 100 or more prickles in view at one time; leaves thick.
                7. Burs villous-hirsute, 6–8 mm thick . . . . 7. *X. italicum*
                7. Burs pilose, 8–12 mm thick . . . . 8. *X. pensylvanicum*

1. **Xanthium spinosum** L. Sp. Pl. 2:987. 1753. Fig. 6.

Annual herbs from fibrous roots; stems erect, branched, to 1 m tall, pubescent and usually scabrous; leaves alternate, narrowly lanceolate to ovate, pinnately 3-, 5-, or 7-lobed, acuminate at the apex, tapering to the petiolate base to 8 cm long, to 3.5 cm wide, thin, strigose on the upper surface, white-hairy on the lower surface, with yellow axillary spines up to 2 cm long; burs oblongoid, more than twice as long as broad, 10–13 mm long, 5–7 mm thick, brown, with or without a short beak, the body somewhat pubescent, with 80–100 prickles in one view, the glabrous prickles 2.5–4.0 mm long, uncinate.

Common Name: Spiny cocklebur.
Habitat: Disturbed wet areas.
Range: Native to tropical America; rarely adventive in Illinois.
Illinois Distribution: Known only from Alexander, Cook, and Pulaski counties.

This is the only *Xanthium* with axillary spines and leaves that are pinnately lobed and tapered to the base.

Xanthium spinosum fruits from April to October.

species	bur color	bur length	bur thickness	bur shape	length/ ratio	beak length	bur hairiness	prickles in 1 view	prickle length	prickle tip
chinense	tawny	20–25	10–18	ellipse	2× +	3–6	glabrous	100–150	4–7	straight
speciosum	tawny	30–40	15–20	ovoid	2× –	6–11	hispid	200+	5–10	uncinate
globosum	light brown	13–20	10–18	ovoid to globose	2× –	3–4	glabrous	50–80	4–6	straight
inflexum	tawny	20–25	10–15	oblong	2× +	5–7	hirtellous	50–80	4.5–10	uncinate
chasei	brown	13–16	6–9	ovoid to globose	2× –	4–5	glabrous	100–200	2–3.5	straight
italicum	tawny	17–30	6–8	oblong	2× +	5–7	villous-hirsute	100+	4–7	uncinate
pensylvanicum	pale brown	20–25	8–12	oblong	2× +	3–6	pilosulous	200+	4–7	uncinate

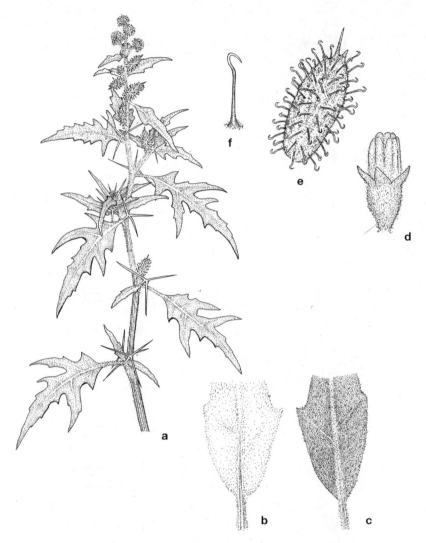

6. *Xanthium spinosum* (Spiny cocklebur).

a. Upper part of plant.
b, c. Leaf.
d. Staminate flower.
e. Fruiting bur.
f. Prickle of bur.

2. **Xanthium speciosum** Kearney, Bull. Torrey Club 24:574. 1897. Fig. 7 (a, b).

Annual herbs from fibrous roots; stems erect, branched, to 1.5 m tall, pubescent or sometimes glabrous, especially near the summit of the stem; leaves alternate, deltoid, ovate, shallowly lobed, acute at the apex, rounded at the petiolate base, to 15 (–18) cm long, nearly as wide, thick, dentate, scabrous on both surfaces; burs ovoid, less than twice as long as broad, 30–40 mm long, 15–20 mm thick, tawny, the beaks 6–11 mm long, the body hispid, with 200 or more prickles in one view, the prickles 5–10 mm long, uncinate.

Common Name: Great clotbur.
Habitat: Low ground, mostly along roads.
Range: Minnesota to South Dakota, south to Texas and Florida; Mexico.
Illinois Distribution: Scattered in Illinois, but not common.

This cocklebur is very distinctive because of its longer burs and its more slender prickles that number more than two hundred in one view of each bur. The beaks at the tip of each bur are longer than in any other species of *Xanthium*.

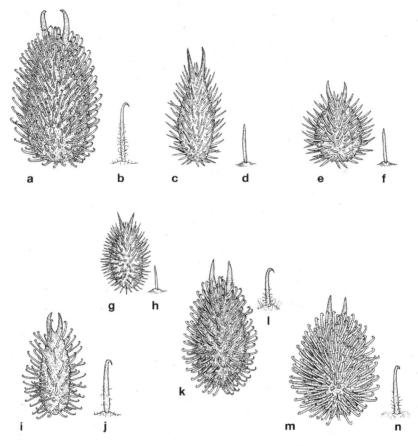

7. *Xanthium speciosum* (Great clotbur).
a. Fruiting bur.
b. Prickle of bur. var. *chinense* (Mexican cocklebur).
c. Fruiting bur.
d. Prickle of bur. var. *globosum* (Subspherical cocklebur).

e. Fruiting bur.
f. Prickle of bur. var. *chasei* (Chase's cocklebur).
g. Fruiting bur.
h. Prickle of bur. var. *inflexum* (Missouri cocklebur).
i. Fruiting bur.

j. Prickle of bur. var. *italicum* (Common cocklebur).
k. Fruiting bur.
l. Prickle of bur. var. *pensylvanicum* (Pennsylvania cocklebur).
m. Fruiting bur.
n. Prickle of bur.

This uncommon species is scattered in Illinois.

Xanthium speciosum fruits in September and October.

3. **Xanthium chinense** Mill. Gard. Dict., ed. 8, no. 4. 1768. Fig. 7 (c, d).
Xanthium glabratum Britt. Man. 912. 1901.

Annual herbs from fibrous roots; stems erect, branched, to 2 m tall, scabrous;
leaves alternate, ovate to deltoid, acute at the apex, cordate or truncate at the
petiolate base, to 18 cm long, nearly as wide, thin, dentate, sometimes shallowly
lobed, scabrous on both surfaces; burs usually ellipsoid, mostly more than twice as
long as broad, 20–25 mm long, 10–18 mm thick, tawny, the beaks 3–6 mm long,
the body glabrous, with 100–150 prickles on one view, the prickles 4–7 mm long,
straight or nearly so.

Common Name: Mexican cocklebur.
Habitat: Wet fields and along roads.
Range: Quebec to Minnesota, south to Nebraska, Texas, and Florida; Mexico.
Illinois Distribution: Scattered and occasional throughout the state.

Despite the scientific epithet, this plant was not discovered in China but in Mexico.
It is scattered in most of Illinois.

The burs are shiny, whereas the burs of all other species of *Xanthium* in Illinois
are dull.

The glabrous bodies of the burs with 100–150 straight prickles in one view are
distinctive. *Xanthium chasei* is similar with glabrous burs and 100–150 prickles in
one view, but differs by its smaller burs and very thick leaves.

Xanthium chinense fruits from August to October.

4. **Xanthium globosum** Shull, Bot. Gaz. 59:482–483. 1915. Fig. 7 (e, f).

Annual herbs from fibrous roots; stems erect, branched, to 1 m tall, scabrous;
leaves alternate, deltoid-ovate, acute at the apex, cordate at the petiolate base, to
15 cm long, nearly as wide, thin, dentate to shallowly lobed, scabrous on both surfaces; burs ovoid to subglobose, not twice as long as broad, 13–20 mm long, 10–18
mm thick, light brown, the beaks 3–4 mm long, the body glabrous, with 50–80
prickles in one view, the smooth prickles 4–6 mm long, straight.

Common Name: Subspherical cocklebur.
Habitat: Low, wet fields.
Range: Illinois and Kentucky to Kansas.
Illinois Distribution: Apparently restricted to the southern fourth of the state.

This species, with a very restricted overall range, has burs that are scarcely longer
than broad and often nearly spherical. *Xanthium chasei* sometimes has nearly
spherical burs, but the leaves of *X. chasei* are very thick and the burs have more
prickles in one view.

Xanthium globosum fruits in September and October.

5. **Xanthium chasei** Fern. Rhodora 48:66–74, pl. 1013. 1946. Fig. 7 (g, h).

Annual herbs from fibrous roots; stems erect, unbranched, to 60 cm tall, scabrous; leaves alternate, deltoid, acute at the apex, cordate at the petiolate base, to 10 cm long, to 8 cm wide, very thick, dentate, harshly scabrous; burs subglobose to ovoid, up to twice as long as broad, 13–16 mm long, 6–9 mm thick, brown, the beaks 4–5 mm long, the body glabrous, with 100–200 prickles in one view, the prickles 2.0–3.5 mm long, straight.

Common Name: Chase's cocklebur.
Habitat: Wet fields, along rivers.
Range: Known only from central Illinois.
Illinois Distribution: Known only from central Illinois, mostly along the Illinois River.

This species was first collected by Virginius Chase in Illinois in 1919, and is still known only from a small area in central Illinois. It occurs in wet bottomland fields.

Xanthium chasei is distinguished by its glabrous burs 13–16 mm long and 6–9 mm thick. Its straight prickles, which number 100–200 in one view, are 2.0–3.5 mm long. The leaves are very thick.

The fruits mature in September and October.

6. **Xanthium inflexum** Mack. & Bush, Rep. Mo. Bot. Gard. 16:606. 1905. Fig. 7 (i, j).

Annual herbs from fibrous roots; stems erect, branched, to 1.5 m tall, usually scabrous above with papillae; leaves alternate, deltoid-ovate, acuminate at the apex, rounded or subcordate at the petiolate base, to 15 cm long, about as wide, crenate, usually shallowly lobed, thin, scabrous on both surfaces; burs oblongoid, more than twice as long as broad, 20–25 mm long, 10–15 mm thick, stramineous or tawny, the beaks 5–7 mm long, the body hirtellous, with 5–80 prickles in view at one time, the sometimes pubescent prickles 4.5–10.0 mm long, uncinate.

Common Name: Missouri cocklebur.
Habitat: Wet bottomlands.
Range: Illinois and Missouri to Arkansas.
Illinois Distribution: Confined to the southern tip of the state.

This species is recognized by its elongated burs 20–25 mm long with only 50–80 uncinate prickles in view at one time. The body of the burs is usually hirtellous.

Xanthium inflexum fruits in September and October.

7. **Xanthium italicum** Moretti, Giorn. Fis., ser. 2, 5:326. 1822. Figs. 7 (k, l), 8.
Xanthium commune Britt. Man. 912. 1901.

Annual herbs from fibrous roots; stems erect, usually branched, to 80 cm tall, scabrous; leaves alternate, ovate, acute at the apex, usually cordate at the petiolate

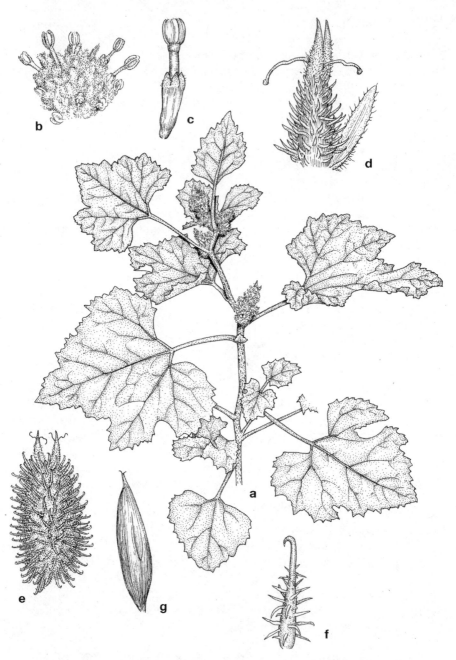

8. *Xanthium italicum* (Common cocklebur).

a. Upper part of plant.

b. Involucre of disc flowers.

c. Disc flower.

d. Pistillate flower with phyllary.

e. Fruiting bur.

f. Prickle of bur.

g. Seed.

base, sometimes shallowly lobed, to 18 cm long, nearly as wide, thick, dentate, harshly scabrous on both surfaces; burs oblongoid, more than twice as long as broad, 17–30 mm long, 6–8 mm thick, tawny, the beaks 5–7 mm long, the body villous-hirsute, with more than 100 prickles in view at one time, the prickles 4–7 mm long, hirsute, uncinate.

Common Name: Common cocklebur.
Habitat: Low ground.
Range: Quebec to Saskatchewan, south to Texas and Florida.
Illinois Distribution: Common throughout the state.

This species is probably the most common cocklebur throughout the state of Illinois.

The major distinguishing feature of *X. italicum* is the villous-hirsute body of the bur. It is one of four species with thick leaves.

Xanthium italicum fruits in September and October.

8. **Xanthium pensylvanicum** Wallr. Beitr. Bot. 1:236. 1842. Fig. 7 (m, n).

Annual herb from fibrous roots; stems erect, branched, to 80 cm tall, scabrous but sometimes smooth near the base; leaves alternate, ovate, acute to acuminate at the apex, rounded or subcordate at the petiolate base, to 15 cm long, nearly as wide, thick, dentate or serrate, somewhat scabrous; burs oblongoid, more than twice as long as broad, 20–25 mm long, 8–12 mm thick, pale brown, the beaks 3–6 mm long, the body pilosulous, with more than 200 prickles in view at one time, the prickles 4–7 mm long, hispid, uncinate.

Common Name: Pennsylvania cocklebur.
Habitat: Low ground in fields.
Range: Quebec to California, east to Texas and Florida.
Illinois Distribution: Scattered throughout most of Illinois.

This species is similar in appearance to *X. italicum*, but the burs are thicker, pale brown, have more prickles in view at one time, and have a pilosulous body.

Xanthium pensylvanicum fruits in September and October.

32. **Parthenium** L.—Feverfew

Annual or perennial herbs; stems erect, branched; leaves basal and/or alternate; basal leaves when present petiolate; cauline leaves toothed or pinnately divided, pubescent; heads radiate, usually in corymbs or cymes; involucres hemispheric; phyllaries 10–15, in 2 series; receptacle flat to conic, with paleae; ray flowers pistillate, fertile, whitish; disc flowers 12–60, staminate, the corolla 5-lobed, whitish; cypselae obovoid, usually shed with one phyllary attached; pappus absent, or represented by minute projections.

There are seventeen species in the genus, all native to the New World.

1. Leaves pinnately or bipinnately divided . . . . 1. *P. hysterophorus*

1. Leaves serrate or dentate.
    2. Stems and leaves glabrous or variously pubescent but not hispid . . . 2. *P. integrifolium*
    2. Stems and leaves hispid . . . . 3. *P. hispidum*

   1. **Parthenium hysterophorus** L. Sp. Pl. 2:988. 1753. Fig. 9.
    Annual herbs from fibrous roots; stems erect, branched, to 80 cm tall, strigose to villous; leaves cauline, alternate, ovate to oblong, pinnately or bipinnately

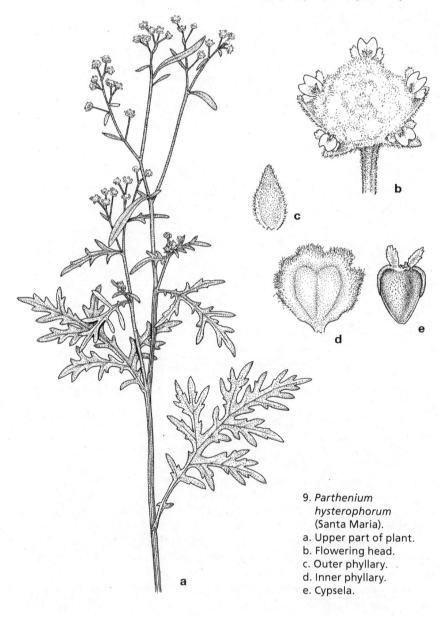

9. *Parthenium hysterophorum* (Santa Maria).
a. Upper part of plant.
b. Flowering head.
c. Outer phyllary.
d. Inner phyllary.
e. Cypsela.

lobed, to 5 cm long, to 2.5 cm wide thin, scabrous on both surfaces, glandular; heads numerous, mostly in cymes, rather obscurely radiate; involucres hemispheric; phyllaries about 10, in 2 series, the outer lance-elliptic, 2–4 mm long, the inner ovate to suborbicular, 2.5–4.0 mm long; receptacle flat, paleate; ray flowers 5 per head, pistillate, fertile, white, more or less orbicular, up to 1 mm long; disc flowers 30–50, staminate, gray-white, the corolla 5-lobed; cypselae obovoid, 1.5–3.0 mm long, glabrous; pappus of two ovate projections about 1 mm long.

Common Name: Santa Maria.
Habitat: Disturbed soil (in Illinois).
Range: Native to tropical America; adventive from Connecticut to Minnesota, south to New Mexico, Texas, and Florida.
Illinois Distribution: Cook County; not seen since 1891.

This non-native is distinguished by its pinnatifid or bipinnatifid cauline leaves and its small ray flowers with white rays up to 1 mm long.

Parthenium hysterophorus flowers from August to October.

2. **Parthenium integrifolium** L. Sp. Pl. 2:988. 1753. Fig. 10.
Perennial herbs from tuberous thickened roots, not forming stolons; stems erect, rather stout, branched above, to 1 m tall, glabrous or puberulent; basal leaves ovate to obovate, acute at the apex, rounded at the petiolate base, to 30 cm long, to 15 cm wide, thick, scabrous, serrate or crenate; cauline leaves alternate, lanceolate, acute to short-acuminate at the apex, rounded at the base, the lowest on short petioles, the uppermost sessile, serrate or dentate, scabrous or hispidulous; heads numerous in a terminal corymb up to 25 cm across, radiate; involucre conic, 3.5–6.0 mm high; phyllaries about 10, in 2 series, the outer oblong to ovate, 3–5 mm long, gray-pubescent, the inner orbicular, 4–6 mm long, gray-pubescent; receptacle more or less flat, paleate; ray flowers 5 per head, pistillate, fertile, white, oblong to orbicular, 1–2 mm long; disc flowers up to 30 (–35) per head, staminate, whitish, the corolla 5-lobed; cypselae obovoid, glabrous, 3–4 mm long; pappus of 2 delicate projections about 0.5 mm long.

Common Name: American feverfew; wild quinine.
Habitat: Prairies, dry woods, rocky glades, black oak savannas.
Range: Connecticut to Minnesota, south to Texas and Florida.
Illinois Distribution: Common throughout the state.

This species is similar to *P. hispidum*, differing by larger stature, larger flowering heads, tuberous-thickened, non-stoloniferous rootstocks, and harshly pubescent stems and leaves.

It is a common inhabitant of prairies, dry woods, and rocky glades.

Parthenium integrifolium flowers from June to September.

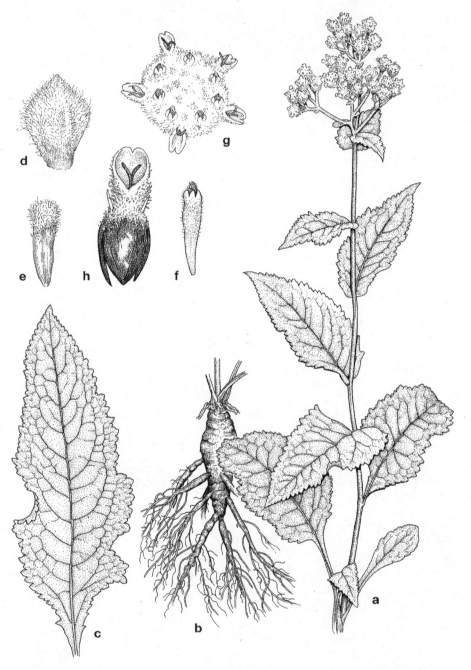

10. *Parthenium integrifolium* (American feverfew).

a. Habit.
b. Lower part of plant.
c. Leaf.
d. Outer phyllary.

e. Inner phyllary.
f. Unopened flower.
g. Flowering head.
h. Ray flower.

3. **Parthenium hispidum** Raf. New Fl. N. Am. 2:35. 1836. Fig. 11.

Perennial herbs from non-tuberous thickened rootstocks, with slender, creeping stolons; stems erect, rarely sprawling, branched, to 75 cm tall, hispid with spreading hairs; basal leaves, if present, ovate, acute at the apex, rounded at the petiolate base, to 22 cm long, to 10 cm wide, thick, harshly scabrous and hispid, serrate; cauline leaves alternate, lance-ovate, acute at the apex, rounded at the base, the lowest petiolate, the uppermost sessile, serrate, harshly scabrous, hispid;

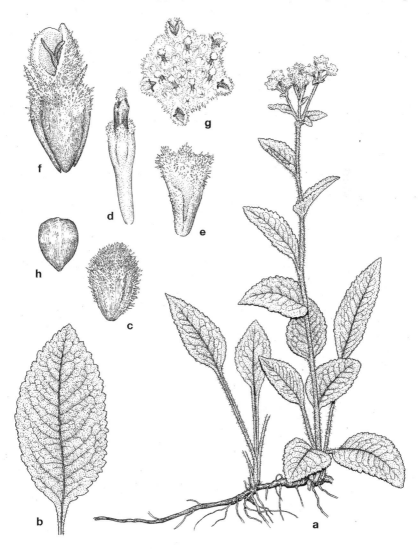

11. *Parthenium hispidum* (Hispid fevervew).
a. Habit.
b. Leaf.
c. Outer phyllary.
d. Disc flower.
e. Inner phyllary.
f. Ray flower.
g. Flowering head.
h. Cypsela.

heads several to numerous in a terminal corymb up to 15 cm across, radiate; involucre conical, 4–7 mm high, 6–10 mm wide; phyllaries about 10, in 2 series, the outer narrowly ovate, 3–5 mm long, hispidulous, the inner orbicular, 4–6 mm long, hispidulous; receptacle more or less flat, paleate; ray flowers 5 per head, pistillate, fertile, white, orbicular, 1–2 mm long; disc flowers about 30 per head, staminate, white, the corolla 5-lobed; cypselae obovoid, glabrous, 3–4 mm long; pappus of 2 projections up to 0.5 mm long.

Common Name: Hispid feverfew.
Habitat: Rocky glades.
Range: Illinois to Kansas, south to Texas and Arkansas.
Illinois Distribution: Apparently confined to the southern half of the state.

Plants with very extensive, creeping stolons have been called *P. repens*, but recognition of this binomial does not appear to be justified because all specimens of *P. hispidum* have creeping stolons, either long or short.

The corymbs of *P. hispidum* are smaller than those of *P. integrifolium*.

Parthenium hispidum flowers from May to early August, usually a little earlier than *P. integrifolium*.

33. **Iva** L.—Marsh Elder

Annual herbs (in Illinois); stems erect, branched; leaves cauline, opposite except sometimes for the uppermost, the larger usually 3-nerved, simple, toothed, unlobed; heads discoid, numerous, in racemes or spikes, each head subtended by a linear bract; involucre hemispheric; phyllaries 3–10, the outer leaf-like; receptacle usually flat, the paleae enclosing the florets; marginal flowers pistillate, tubular; central flowers of disc up to 20, funnelform, staminate, whitish, with 5 reflexed lobes; cypselae obovoid, more or less scabrous; pappus absent.

There are nine species of *Iva*, all native to the New World. Only one species occurs in Illinois since *I. xanthifolia*, which used to be in the genus, is now in *Cyclachaena*.

1. **Iva annua** L. Sp. Pl. 2:288. 1753. Fig. 12.
Iva ciliata Willd. Sp. Pl. 3:2386. 1804.

Coarse annuals from fibrous roots; stems erect, branched or unbranched, to 2 m tall, scabrous, hispid; leaves cauline, all but sometimes the uppermost opposite, ovate, acute to acuminate at the apex, tapering to the base, the lowest usually petiolate, the upper sometimes sessile, to 7 cm long, to 4.5 cm wide, serrate, scabrous on both surfaces, glandular, the petioles, when present, usually ciliate; heads discoid, numerous, in spikes; involucre hemispheric, 3–5 mm wide; phyllaries 3–10, the outer bract-like and longer than the head, hispid, 6–15 mm long; receptacle hemispheric, paleate; marginal flowers 3–5, pistillate, to 1 mm long; central flowers 8–15 per head, staminate, the corollas 2.0–2.5 mm long; cypselae obovoid, glabrous or nearly so, 2–3 mm long; pappus absent.

12. *Iva annua* (Marsh elder).
a. Upper part of plant.
b. Leaf.
c. Phyllary.
d. Raceme of staminate
   flowers.
e. Staminate flower with
   phyllary.
f. Staminate flower.
g. Pistillate flower.
h. Cypsela.

Common Name: Marsh elder; sumpweed.
Habitat: Moist, open ground.
Range: Maine to North Dakota, south to New Mexico and Florida.
Illinois Distribution: Common in the southern third of Illinois, less common
    northward, and possibly adventive in the northeastern counties.

This species is the same as Willdenow's *I. ciliata*, the binomial by which it was
called for many years.

Iva annua strongly resembles specimens of *Ambrosia trifida* that have unlobed
leaves. However, each head in *Iva* is subtended by a foliaceous bract, while each
head of *Ambrosia trifida* lacks these conspicuous bracts.

Iva annua flowers from August to October.

34. **Cyclachaena** Fresen.—Burweed

Annual herbs; stems erect; leaves cauline, opposite, sessile, sometimes lobed;
heads disciform, borne in panicles, sessile or nearly so; involucre mostly hemispheric; phyllaries 10–12 in 2 series; receptacle convex, paleate; pistillate flowers
5, with or without a corolla; staminate flowers 5–15, whitish, 5-lobed; cypselae
obovoid; pappus absent.

Only the following species comprises the genus. The only species has usually
been placed in *Iva*, but it differs from *Iva* by its paniculate inflorescence near absence of pistillate corollas.

1. **Cyclachlaena xanthifolia** (Nutt.) Fresen. Ind. Sem. Hort. Franc. 4. 1836. Fig. 13.
Iva xanthifolia Nutt. Gen. N. Am. Pl. 2:185. 1818.

Annual herbs from fibrous roots; stems erect, branched, to 2 m tall, pubescent in
the upper half, often glabrous below; leaves cauline, opposite, simple, ovate, acuminate at the apex, rounded at the base or less commonly cordate, to 15 cm long, to 12
cm wide, 3- or 5-nerved, serrate, densely tomentellous, the petioles to 10 cm long;
heads in panicles, sessile, disciform; involucre hemispheric, 2–3 mm across; phyllaries
10 or 12, in 2 series, the outer pubescent, the inner glabrous or nearly so; receptacle
convex, the paleae 2.0–2.5 mm long; pistillate flowers 5 per head, the corolla, if present, minute and whitish; staminate flowers 5–15, 2.0–2.5 mm long, the corolla whitish, 5-lobed; cypselae oblongoid, 2–3 mm long, sparsely pubescent; pappus absent.

Common Name: Burweed.
Habitat: Moist, disturbed areas.
Range: Nova Scotia to British Columbia, south to California, Texas, Kentucky, and
    Virginia.
Illinois Distribution: Occasional in the northern half of the state.

Until recently, most botanists have placed this species in *Iva*. It differs by its flowering heads borne in panicles, rather than racemes or spikes.

Cyclachaena xanthifolia flowers from July to September.

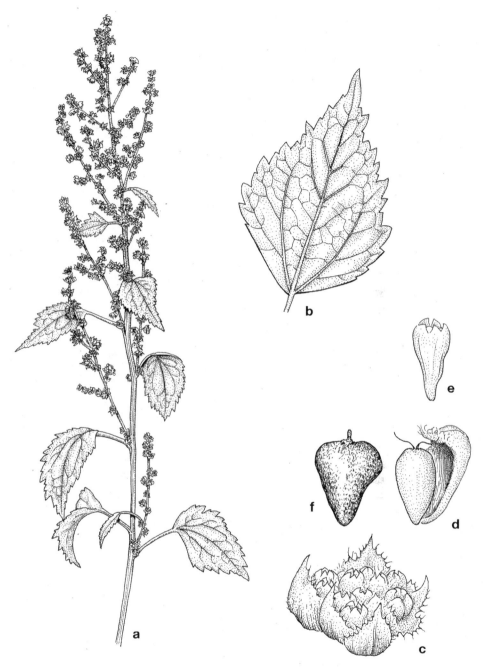

13. *Cyclachaena xanthifolia*
    (Burweed).
a. Upper part of plant.
b. Leaf.

c. Cluster of staminate
    flowers.
d. Pistillate flower and
    phyllary.

e. Staminate flower.
f. Cypsela.

Subtribe Melanpodiinae Less.

Perennial herbs (in Illinois); stems erect; leaves cauline, opposite, simple, usually lobed; heads radiate, borne singly or in corymbs; involucre hemispheric (in
Illinois); phyllaries 8–25 in 2 series; receptacle flat or conic, paleate; ray flowers
3–20, pistillate, fertile, yellow; disc flowers few to numerous, staminate, yellow, the
corolla 5-lobed; cypselae obovoid, more or less flattened; pappus absent.

This subtribe consists of five genera and about eighty species, mostly in the New
World tropics. Only *Smallanthus* occurs in Illinois.

35. **Smallanthus** Mack. ex Small—Bear's-foot

Perennial herbs; stems erect; leaves cauline, opposite, simple, palmately lobed, petiolate; heads few or in a corymb, radiate; involucre hemispheric; phyllaries 12–20,
in 2 series; receptacle flat or convex, paleate; ray flowers 7–15, pistillate, fertile,
yellow; disc flowers 30 or more, staminate, yellow, the corolla 5-lobed; cypselae
oblongoid, more or less flattened, striate; pappus absent.

Smallanthus consists of about twenty New World species, only one of which is in
the United States. This species traditionally has been placed in *Polymnia*.

1. **Smallanthus uvedalia** (L.) Mack. ex Small, Man. S. E. Flora 1509. 1933. Fig.
14.
Osteosperma uvedalia L. Sp. Pl. 2:923. 1753.
Polymnia uvedalia (L.) L. Sp. Pl., ed. 2, 1303. 1763.

Perennial herbs from rhizomes; stems erect, branched, stout, to 3 m tall, rough-
pubescent; leaves cauline, opposite, simple, aromatic, broadly ovate, usually with
4 angular palmate lobes, acute at the apex, more or less truncate at the base, to 50
cm long, nearly as wide, the lobes toothed, the lower leaves with a winged petiole,
the upper leaves sessile and often clasping, pubescent on both surfaces; heads few
in a corymb; involucre hemispheric, 8–15 mm across; phyllaries usually 12–20 in
2 series, the outer 4–6 very large, ovate to orbicular, the inner usually the same
number as the rays, lanceolate; receptacle flat to convex, paleate; ray flowers
10–15, pistillate, fertile, yellow, narrowly oblong, up to 30 mm long; disc flowers
30 or more, staminate, yellow, the corolla 5-lobed; cypselae oblongoid, usually flattened, 5–6 mm long, with 30–40 fine striations; pappus absent.

Common Name: Bear's-foot; yellow-flowered leaf-cup.
Habitat: Rich woods.
Range: New York to Wisconsin, south to Kansas, Texas, and Florida.
Illinois Distribution: Occasional in southern Illinois, extending northward to St.
Clair County; also Vermilion County.

This species, long considered to be in *Polymnia*, differs by its palmately lobed
leaves, more ray flowers, yellow rays, and striated cypselae.

The large leaves are somewhat aromatic when crushed.

Smallanthus uvedalia is found in rich woods. It flowers from July to September.

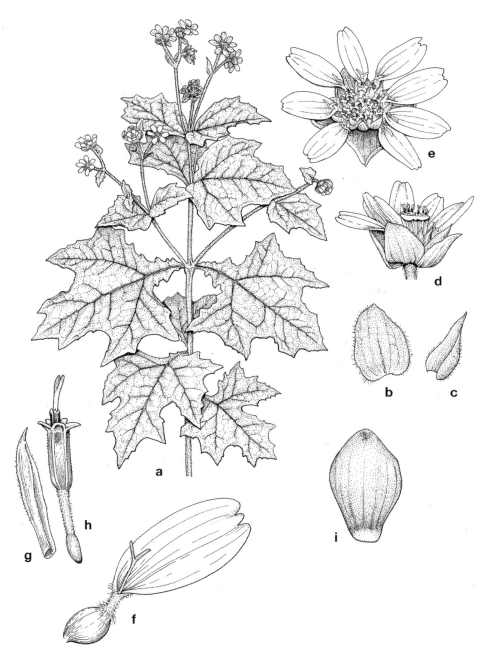

14. *Smallanthus
    uvedalia* (Bear's-
    foot).
a. Upper part of plant.
b. Outer phyllary.

c. Inner phyllary.
d. Flowering head,
   side view.
e. Flowering head,
   face view.

f. Ray flower.
g. Palea.
h. Disc flower.
i. Cypsela.

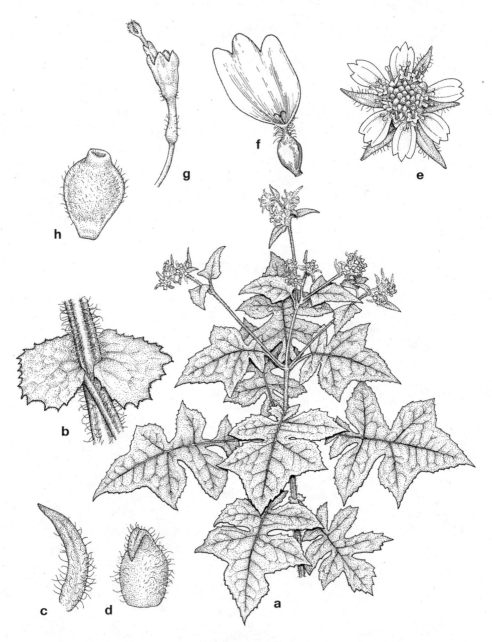

15. *Polymnia*
   canadensis
   (Leaf-cup).
a. Upper part of plant.

b. Leaf.
c. Inner phyllary.
d. Outer phyllary.
e. Flowering head.

f. Ray flower.
g. Disc flower.
h. Cypsela.

Subtribe Polymniinae H. Robins.

Perennial herbs (in Illinois); stems erect, branched; leaves cauline, opposite, simple, shallowly pinnately lobed; heads radiate, borne in corymbs; involucre hemispheric; phyllaries 6–21, in 2 series; receptacle flat or convex, paleate; ray flowers 2–6, pistillate, fertile, whitish or yellowish; disc flowers 12 or more, staminate, the corolla pale yellow, 5-lobed; cypselae obovoid, angled or ribbed, smooth or nearly so; pappus absent.

Only one genus and three species, all in the eastern United States, comprise this subtribe.

36. Polymnia L.—Leaf-cup

Perennial herbs; stems erect, branched; leaves cauline, simple, opposite, shallowly pinnately lobed; heads radiate, borne in corymbs; involucre hemispheric; phyllaries 6–21, in 2 series; receptacle flat or convex, paleate; ray flowers 2–6, pistillate, fertile, whitish or yellowish; disc flowers 12 or more, staminate, the corolla pale yellow, 5-lobed; cypselae obovoid, angled or ribbed, smooth or nearly so; pappus absent.

Only the following species occurs in Illinois.

1. **Polymnia canadensis** L. Sp. Pl. 2:926. 1753. Fig. 15.
Polymnia canadensis L. var. *discoidea* Gray, Man. Bot., ed. 3:248. 1857.
Polymnia canadensis L. var. *radiata* Gray, Syn. Fl. N. Am. 1, part 2:238. 1884.
Polymnia radiata (Gray) Small, Fl. S. E. U. S. 1239. 1903.

Perennial herbs from rhizomes; stems erect, branched above, to 1.5 m tall, viscid-pubescent; leaves cauline, opposite, simple, aromatic, ovate to hastate, acute at the apex, rounded to truncate at the base, to 30 cm long, to 20 cm wide, 5- or 7-pinnately lobed, viscid-pubescent, on winged petioles to 10 cm long; heads usually radiate, 2–6 in a terminal corymb; involucre hemispheric, 8–15 mm across; phyllaries 6–21, in 2 series, the outer large, narrowly ovate to ovate, the inner narrower; receptacle flat to convex, paleate; ray flowers 2–6, pistillate, fertile, often 0.2–1.0 cm long, usually with 3 shallow notches at the tip, rarely absent, whitish; disc flowers 12 or more, staminate, pale yellow, the corolla 5-lobed; cypselae obovoid, 3-angled, more or less flattened, with 3 ribs, 3–4 mm long; pappus absent.

Common Name: Leaf-cup.
Habitat: Moist or dry woods, shaded limestone slopes, forested fens, seeps.
Range: New York to Minnesota, south to Oklahoma and Georgia.
Illinois Distribution: Common throughout the state.

Typical *P. canadensis* is rayless or has rays less than 0.5 mm long. Many Illinois plants have rays about 1 cm long. These have sometimes been segregated as var. *radiata*, or even as a separate species called *P. radiata*. Since there is intergradation on the same plant from rayless to rays up to 1 cm long, I have elected not to recognize this variety.

Polymnia canadensis flowers from June to November.

Subtribe Milleriinae Benth. & Hook.

Annual herbs (in Illinois) from fibrous roots; stems erect, branched; leaves cauline, opposite (in Illinois), simple, glandular; heads radiate, in corymbs; involucre usually campanulate; phyllaries usually 10–12, in 2 series; receptacle conic to convex, paleate; ray flowers pistillate, fertile, yellow; disc flowers numerous, bisexual, fertile, yellow, the corolla 5-lobed; cypselae obovoid or angular, somewhat flattened; pappus absent.

This subtribe contains six genera and thirty-two species, none native to the United States and only one introduced in this country.

37. **Guizotia** Cass. in F. Cuvier—Ramtilla

Annuals (in Illinois); stems erect, branched; leaves cauline, mostly opposite, simple; heads radiate, 1 to few, in corymbs; involucre campanulate; phyllaries 10–12, in 2 series; receptacle conic to convex, paleate; ray flowers (6–) 8–12 (–18), pistillate, fertile, yellow; disc flowers numerous, bisexual, fertile, yellow, the corolla 5-lobed; cypselae obovoid or angular, more or less compressed; pappus absent.

Guizotia contains six African species. The following from Illinois represents an escape from cultivation.

1. **Guizotia abyssinica** (L.f.) Cass. in F. Cuvier, Dict. Sci. Nat. ed. 2, 59:248. 1829. Fig. 16.
Polymnia abyssinica L.f. Suppl. Pl. 383. 1782.

Annual from fibrous roots; stems erect, branched, to 1.5 m tall, glabrous, papillose; leaves cauline, opposite except sometimes for the uppermost, elliptic to oblong-lanceolate, acute at the apex, tapering or rounded at the sessile and sometimes clasping base, to 15 cm long, to 6 cm wide, irregularly serrate, glabrous, glandular; heads radiate, 2–3 cm across, 1 to few, in a corymb or cyme; involucre campanulate, 8–20 mm across; phyllaries 10–12, in 2 series, the outer more or less foliaceous, the inner scarious; receptacle conic to convex, the paleae 5–7 mm long; ray flowers (6–) 8–12, pistillate, fertile, yellow, 10–15 mm long, with 3 notches at the apex; disc flowers numerous, bisexual, yellow, the tube tomentose, the corolla 5-lobed; cypselae obovoid, flattened, 3- or 4-angular, 4–5 mm long, glabrous; pappus absent.

Common Name: Ramtilla.
Habitat: Disturbed soil.
Range: Native to Africa; adventive from several states east of the Mississippi River; also Arkansas, California, and Kansas.
Illinois Distribution: Reported without precise locality by Rydberg, based on a specimen in the New York Botanical Garden Herbarium.

This is a pretty species with yellow flowering heads up to 3 cm across. Both the ray flowers and the disc flowers are fertile. The flattened cypselae from the ray flowers are 3-angled, while those from the disc flowers are 4-angled.

Guizotia abyssinica flowers in September and October.

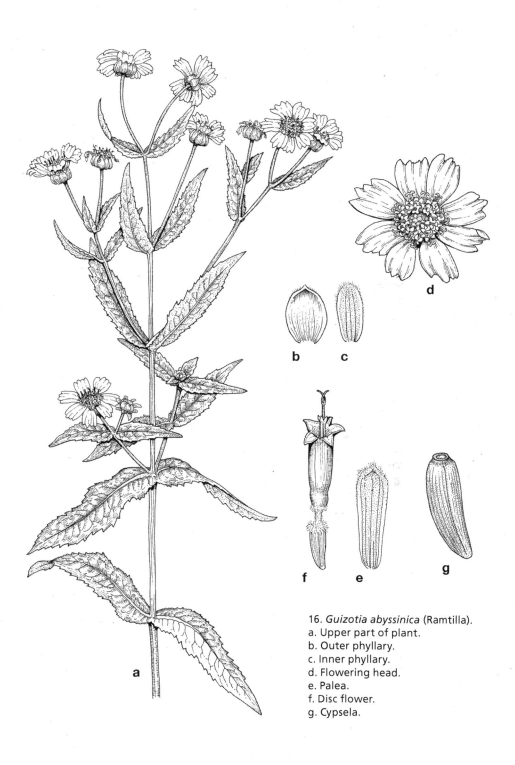

16. *Guizotia abyssinica* (Ramtilla).
a. Upper part of plant.
b. Outer phyllary.
c. Inner phyllary.
d. Flowering head.
e. Palea.
f. Disc flower.
g. Cypsela.

Subtribe Rudbeckiinae H. Robins.

Annual or perennial herbs; stems erect, branched or unbranched; leaves basal and cauline, simple, the cauline alternate, sometimes lobed or pinnately cleft; heads radiate, borne in racemes, panicles, or solitary; involucre hemispheric or rotate; phyllaries 14–30, in 1 to several series; receptacle ovoid to conic to columnar, epaleate or, if paleate, the paleae folded; ray flowers up to 25, neutral, yellow, or orange, sometimes reddish; disc flowers numerous, bisexual, fertile, yellow to brown-purple, the corolla 5-lobed; cypselae subterete or angular or flattened; pappus absent, or with a few scales or teeth.

This subtribe includes three genera and about thirty-five species, all in North America, and all three genera occur in Illinois.

Key to the Genera of Subtribe Rudbeckiinae in Illinois
1. Leaves lobed or deeply divided.
    2. Receptacle epaleate; phyllaries in 2 series; rays strongly reflexed. . . . . . . 40. *Ratibida*
    2. Receptacle paleate; phyllaries in several series; rays spreading or only slightly
       reflexed . . . . 39. *Rudbeckia*
1. Leaves neither lobed nor deeply divided.
    3. Rays reflexed; disc long-columnar. . . . . 38. *Dracopis*
    3. Rays spreading, rarely slightly reflexed; disc globose . . . . 39. *Rudbeckia*

38. **Dracopis** Cass.—Clasping Coneflower

Annual; stems erect, branched, glabrous; leaves cauline, simple, alternate, auriculate, glaucous; head solitary, radiate; involucre flat; phyllaries several, in 2 series, spreading to reflexed; receptacle conic, paleate; ray flowers yellow, notched at apex, neutral; disc flowers numerous, bisexual, fertile, brownish, the corolla 5-lobed; cypselae terete, striate; pappus absent.

Only the following species comprises the genus. It is often placed in *Rudbeckia*, but differs by its terete rather than angular cypselae that are completely devoid of any pappus, and by its glabrous, glaucous, auriculate leaves.

1. **Dracopis amplexicaulis** (Vahl) Cass. in DC. Prod. 5:558. 1836. Fig. 17.
Rudbeckia amplexicaulis Vahl, Skr. Nat. Selsk. 2:29. 1793.

Annual herbs from fibrous roots; stems erect, branched, to 75 cm tall, glabrous, glaucous; leaves cauline, alternate, simple, the lower oblong, the upper ovate, to 15 cm long, to 4 cm wide, entire, glabrous, glaucous, 1-nerved, the lower sessile, the upper cordate-clasping; head solitary, radiate, on a glabrous peduncle up to 15 cm long; involucre flat, up to 4 cm across; phyllaries several, in 2 series, linear to lanceolate, to 10 mm long, spreading or reflexed, glabrous, ciliate; receptacle conic, paleate; ray flowers 6–15, yellow, often with a brown base, neutral, to 30 mm long, notched at the apex; disc flowers numerous, bisexual, fertile, brown, the corolla 5-lobed, the disc oblongoid to conic, up to 2 cm long; cypselae terete, 4–6 mm long, striate; pappus absent.

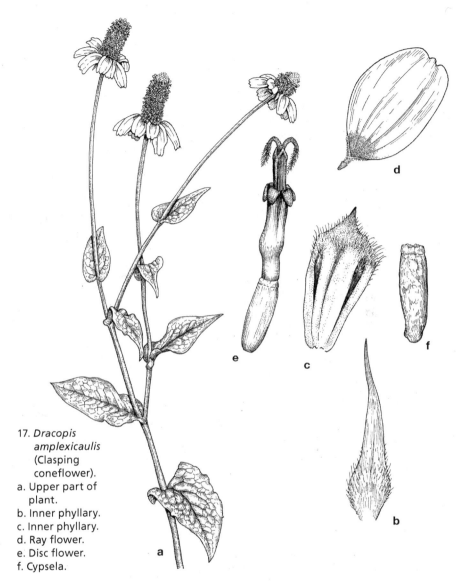

17. *Dracopis amplexicaulis* (Clasping coneflower).
a. Upper part of plant.
b. Inner phyllary.
c. Inner phyllary.
d. Ray flower.
e. Disc flower.
f. Cypsela.

Common Name: Clasping coneflower.

Habitat: Disturbed soil (in Illinois).

Range: South Carolina to Kansas, south to New Mexico and Florida; adventive in Illinois and North Dakota.

Illinois Distribution: Known only from Cook, DuPage, Greene, and Jackson counties.

This handsome species is readily recognized by its glabrous and glaucous clasping leaves and its large yellow flowering heads with a dome-like disc.

This plant is often placed in the genus *Rudbeckia*.
Dracopis amplexicaulis flowers from July to September.

39. **Rudbeckia** L.—Black-eyed Susan

Biennial or perennial herbs, sometimes with rhizomes; stems erect, branched; leaves basal and/or cauline, if cauline, alternate, simple, sometimes few-lobed or pinnatifid, sometimes glandular; heads radiate (in Illinois), solitary or in corymbs or panicles; involucre hemispheric or rotate; phyllaries up to 20, usually in 2 series, subequal; receptacle conic or convex, paleate; ray flowers neutral, yellow, sometimes with a brown or reddish base, spreading or somewhat reflexed; disc flowers numerous, bisexual, fertile, yellow or brown or purple-black, the corolla 5-lobed; cypselae 4-angled, flat at the top; pappus of minute scales or absent.

I recognize about thirty species in *Rudbeckia*, all native to North America and Mexico. I have placed *Rudbeckia amplexicaulis* in the genus *Dracopis*.

Key to the Species of *Rudbeckia* in Illinois

1. Disc greenish yellow; leaves deeply 5- to 9-cleft; plants usually glabrous . . . . 1. *R. laciniata*
1. Disc brown or purple; leaves unlobed or 3- (or 5-) lobed; plants usually pubescent.
  2. Lower cauline leaves 3- (or 5-) lobed or parted.
    3. Plants with basal leafy offshoots; palea of receptacle glabrous; leaves rough-hairy on the lower surface . . . . .2. *R. triloba*
    3. Plants without basal leafy offshoots; palea of receptacle pubescent toward apex; leaves downy-pubescent on the lower surface. . . . . .3. *R. subtomentosa*
  2. Lower cauline leaves toothed or entire, not lobed nor parted.
    4. Pappus absent; style branches elongate, acute.
      5. Heads up to 1.5 cm across; leaves nearly uniform; annual . . . . .4. *R. bicolor*
      5. Heads more than 1.5 cm across; lower leaves larger than upper leaves; perennials.
        6. Basal leaves ovate to broadly elliptic, 4–7 cm wide; cauline leaves ovate to occasionally lanceolate, coarsely toothed, the lowest ones 3–6 cm wide . . . . 5. *R. hirta*
        6. Basal leaves oblanceolate to lanceolate, 1–4 (–5) cm wide; cauline leaves linear-lanceolate to oblanceolate, entire or finely serrate, the lowest ones up to 4 cm wide. . . . . .9. *R. serotina*
    4. Pappus present, consisting of a minute, short crown, commonly toothed on the angles; style branches short, obtuse.
      7. Lobes of disc corollas reflexed; rays 4 cm long or longer . . . . 6. *R. grandiflora*
      7. Lobes of disc corollas ascending; rays 1–4 cm long.
        8. Cauline leaves entire, linear-spatulate . . . . .7. *R. missouriensis*
        8. Cauline leaves toothed, elliptic to ovate.
          9. Basal leaves about three times longer than wide; rays 5–20 (–25) mm long.
            10. Basal leaves 2.0–4.5 cm wide; stems villous-hirsute; paleae ciliate . . . . .8. *R. fulgida*
            10. Basal leaves 1–2 cm wide; stems glabrous or strigose; paleae eciliate. . . . . 10. *R. tenax*

9. Basal leaves not more than twice as long as wide; rays 14–40 mm long.
 11. Upper cauline leaves noticeably smaller than lower cauline leaves.
  12. Rays 25–40 mm long; paleae eciliate . . . . 11. *R. sullivantii*
  12. Rays 15–25 mm long; paleae ciliate . . . . 12. *R. palustris*
 11. Upper cauline leaves not noticeably smaller than lower cauline leaves.
  13. Ray flowers 8–12 per head; basal leaves coarsely dentate . . . .
  . . . . 13. *R. umbrosa*
  13. Ray flowers 12–20 per head; basal leaves crenate or entire.
   14. Stems densely villous-hirsute; basal leaves up to 3.5 cm wide,
   coarsely crenate; phyllaries densely hairy; rays 15–25 mm
   long . . . . .14. *R. deamii*
   14. Stems glabrous or sparsely villous; basal leaves up to 6.5 cm
   wide, finely crenate to entire; phyllaries glabrous or sparsely
   hairy; rays 20–24 mm long . . . . .15. *R. speciosa*

1. **Rudbeckia laciniata** L. Sp. Pl. 2:906. 1753. Fig. 18.
Rudbeckia laciniata L. var. *humilis* Gray, Syn. Fl. N. Am. 1:262. 1884.
Rudbeckia laciniata L. var. *hortensis* Bailey, Man. Cult. Plants 765. 1924.

Perennial herbs from slender rhizomes; stems erect, branched, to 3 m tall, glabrous, rarely somewhat pubescent, sometimes glaucous; leaves basal and cauline, the basal and lower cauline ones often withering at anthesis, pinnately divided into 3, 5, 7, or 9 lobes or leaflets, to 50 cm long, to 25 cm wide, acute to acuminate at the apex, tapering or sometimes subcordate at the base, glabrous or minutely pubescent, with petioles to 10 cm long, the cauline ones entire or 3- or 5- or 7-lobed, acute at the apex, tapering or rounded at the base, to 40 cm long, to 20 cm wide, glabrous or minutely pubescent, sessile or short-petiolate; heads 1 to many, to 8 cm across, in corymbs, radiate; involucre hemispheric; phyllaries up to 15, in 2 series, unequal, ovate to lanceolate, glabrous or pubescent, ciliate, to 2 cm long; receptacle hemispheric, the paleae 4.5–6.0 mm long; ray flowers 6–12, yellow, notched at the apex, elliptic, neutral, to 6 cm long, reflexed; disc flowers numerous, yellow-green, bisexual, fertile, the corolla 3.5–5.0 mm long, 5-lobed, the disc to 2.5 cm across; cypselae 4-angled, flattened, 4.2–6.0 mm long, glabrous; pappus of 4 scales 0.4–0.7 mm long.

Common Name: Golden-glow.
Habitat: Floodplain woods, along streams, calcareous springy places.
Range: Nova Scotia to Manitoba, south to Texas and Florida.
*Illinois Distri*bution: Occasional to common throughout Illinois.

This usually tall species has large yellow flowering heads. Cultivated varieties that are double-flowered are popular ornamentals, but these rarely escape into the wild. They may be called var. *hortensis* Bailey.
 Plants with pubescent stems rarely occur in Illinois.
Rudbeckia laciniata flowers from July to November.

18. *Rudbeckia laciniata* (Golden-glow).

a. Habit.
b. Leaf.
c. Phyllaries.
d. Phyllary.

e. Flowering head.
f. Ray flower.
g. Disc flower.
h. Palea.

2. Rudbeckia triloba L. Sp. Pl. 2:907. 1753. Figs. 19a, 19b.

Perennial herbs from slender rhizomes, with leafy offshoots; stems erect, branched, to 1.5 m tall, usually hispid, rarely glabrous; leaves basal and cauline, the basal ones simple, entire, or 3- or 5-lobed, rarely 7-lobed, ovate, acute to acuminate at the apex, often cordate at the base, to 30 cm long, to 8 cm wide, serrate, rough-hirsute on both faces, with petioles to 10 cm long, the cauline ones ovate to lanceolate, acuminate at the apex, tapering to the base, to 20 cm long, to 4 cm wide, serrate or entire, rough-hirsute on both faces, sessile or short-petiolate; heads few to several, to 4 cm across, in panicles, radiate; involucre hemispheric; phyllaries up to 15, in 2 series, linear to lanceolate to lance-ovate, to 1.5 cm long, strigose, hispid, or spreading-villous, reflexed; receptacle conic to hemispheric, the paleae 5.0–6.5 mm long, glabrous, awn-tipped; ray flowers 6–15, to 2 cm long, neutral, yellow or orange-yellow, often with a maroon area near the base; disc flowers numerous, bisexual, fertile, the corolla yellow-green or purple, 3–4 mm long, the disc brown-purple, ovoid, up to 15 mm across; cypselae flattened, 2.0–2.8 mm long, glabrous; pappus of minute scales up to 0.2 mm long.

Two varieties occur in Illinois:

a. Some of the leaves 3-parted or 3-lobed; phyllaries lanceolate to lance-ovate, strigose to hispid . . . . 2a. *R. triloba* var. *triloba*

a. Some of the leaves 5- or 7-parted or -lobed; phyllaries linear to linear-lanceolate, spreading-villous . . . . 2b. *R. triloba* var. *beadleyi*

2a. Rudbeckia triloba L. var. **triloba** Fig. 19a.

Some of the leaves 3-parted or 3-lobed; phyllaries lanceolate to lance-ovate, strigose to hispid.

Common Name: Three-lobed brown-eyed Susan.
Habitat: Woods, fields, stream banks.
Range: Quebec to Minnesota, south to Texas and Florida; also Colorado and Utah.
Illinois Distribution: Common throughout the state.

This plant usually has several leaves that are 3-parted or 3-lobed. The phyllaries are wider and less pubescent than those of the following variety.

Rudbeckia triloba var. *triloba* occurs in woods and fields, but occasionally may be found along streams. It flowers from June to October.

2b. Rudbeckia triloba L. var. **beadleyi** (Small) Fern. Rhodora 39:458. 1937. Fig. 19b.

Rudbeckia beadleyi Small, Fl. S.E.U.S. 1258. 1903.

Some of the leaves 5-parted or 5-lobed; phyllaries linear to linear-lanceolate, spreading-villous.

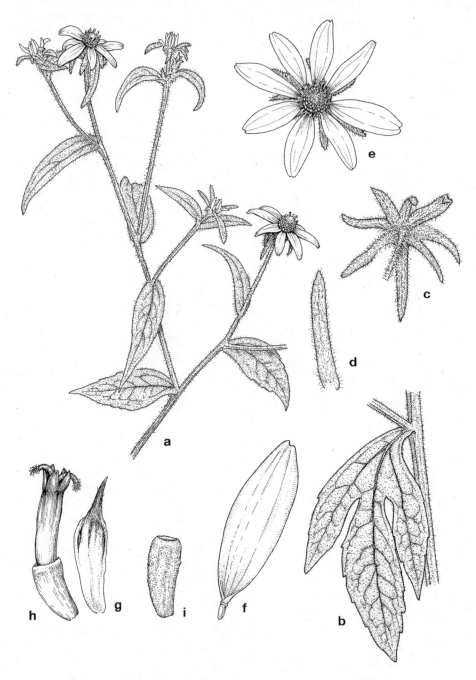

19a. *Rudbeckia triloba* var. *triloba* (Three-lobed brown-eyed Susan).
a. Upper part of plant.
b. Leaf and node.
c. Phyllaries.
d. Phyllary.
e. Flowering head.
f. Ray flower.
g. Palea.
h. Disc flower.
i. Cypsela.

19b. *Rudbeckia triloba* var. *beadleyi* (Five-lobed brown-eyed Susan).

a. Upper part of plant.
b. Cauline leaves.
c. Phyllaries.
d. Flowering head.

e. Ray flower.
f. Palea.
g. Disc flower.
h. Cypsela.

Common Name: Five-lobed brown-eyed Susan
Habitat: Woods, fields.
Range: Maryland to Iowa, south to Oklahoma, Kentucky, and North Carolina.
Illinois Distribution: Scattered in Illinois, but not common.

This variety flowers from June to October.

3. **Rudbeckia subtomentosa** Pursh, Fl. Am. Sept. 2:575. 1814. Fig. 20.
Perennial herbs from stout rhizomes, without leafy offshoots; stems erect, branched, to 2 m tall, densely hirsute, scabrous; leaves basal and cauline, the basal ovate, usually 3-lobed but sometimes 5-lobed, subacute to acute to acuminate at the apex, tapering to the base, to 30 cm long, to 10 cm wide, serrate, densely hirsute and scabrous above, down beneath, the petioles to 10 cm long, the cauline ovate-lanceolate to lanceolate, rounded or tapering to the base, to 20 cm long, to 8 cm wide, serrate, densely hirsute and scabrous above, downy beneath, sessile or nearly so; heads several, to 4 cm across, in corymbs or panicles, radiate; involucre hemispheric; phyllaries up to 15, in 2 series, squarrose, linear-lanceolate, pubescent, to 1.5 cm long; receptacle conic to hemispheric, the paleae 4–6 mm long, acute, with apical pubescence; ray flowers 10–20, to 4 cm long, yellow, sometimes with a darkened base, neutral; disc flowers numerous, brown-purple, perfect, fertile, 3–4 mm long, the corolla, 5-lobed, the disc up to 1.5 cm across, aromatic; cypselae 4-angled, 2.0–3.5 mm long, glabrous or nearly so; pappus of scales 0.1–0.2 mm long.

Common Name: Sweet coneflower; sweet brown-eyed Susan.
Habitat: Open woods, prairies, sand prairies, thickets, stream banks, fens.
Range: New York to Wisconsin, south to Texas, Mississippi, and North Carolina.
Illinois Distribution: Common throughout Illinois.

One of the common names is derived from the faintly sweet aroma of the disc. This species differs from the similar appearing *Rudbeckia triloba* by the absence of basal offshoots, the paleae that are pubescent at the apex, and the downy lower surface of the leaves.

Rudbeckia subtomentosa flowers from July to September.

4. **Rudbeckia bicolor** Nutt. Journ. Nat. Sci. Acad. Phila. 7:81. 1834. Fig. 21.
Annual herbs from a taproot, without basal offshoots; stems erect, branched, to 1.2 m tall, more or less hispid; all leaves cauline, uniform in size, oblong to spatulate, obtuse to subacute at the apex, tapering to the base, 3–6 cm long, 1.5–4.5 cm wide, entire or serrulate, scabrous or occasionally hirsute, usually sessile; heads several in wide-spreading corymbs, up to 2.5 cm across, radiate; involucre hemispheric; phyllaries up to 12, in 2 series, linear-lanceolate, 1–2 cm long; receptacle hemispheric, the paleae 3.5–4.5 mm long, hirsute at least at the tip, ciliate; ray flowers 8–12, 1.5–2.5 cm long, bright yellow, neutral; disc flowers numerous, perfect, fertile, the corolla 2.5–3.5 mm long, 5-lobed, black, the disc up to 1 cm across; cypselae 1.2–2.2 mm long, glabrous or nearly so; pappus absent.

20. *Rudbeckia subtomentosa* (Sweet coneflower).

a. Upper part of plant.
b. Phyllaries.
c. Phyllary.
d. Flowering head.

e. Ray flower.
f. Disc flower with palea.
g. Cypsela.

Common Name: Small-headed black-eyed Susan.
Habitat: Wet areas along streams and in woods.
Range: Indiana to Missouri, south to Texas and Alabama.
Illinois Distribution: Jackson and Union counties.

This species has small bright yellow flowering heads with a small black disc in the center.

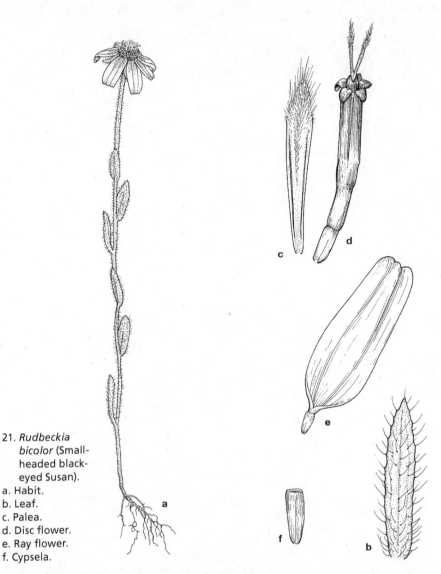

21. *Rudbeckia bicolor* (Small-headed black-eyed Susan).
a. Habit.
b. Leaf.
c. Palea.
d. Disc flower.
e. Ray flower.
f. Cypsela.

In southern Illinois, it occupies wet soil along streams and in woods.

I recognize the plant that Nuttall described. I do not equate it with *Rudbeckia hirta* L. var. *pulcherrima* Farw. that many botanists do.

Rudbeckia bicolor flowers from late June to October.

5. **Rudbeckia hirta** L. Sp. Pl. 2:907. 175. Fig. 22.
Rudbeckia hirta L. var. *pulcherrima* Farw. Rep. Mich. Acad. Sci. 6:209. 1904.

Perennial herbs from taproots; stems erect, branched, to 1 m tall, hispid; basal and cauline leaves present, the basal ovate to broadly elliptic, acute at the apex, tapering or rounded at the base, 4–7 cm wide, serrate, hispid on both surfaces, on petioles to 10 cm long, the cauline ovate or occasionally broadly lanceolate, acute at the apex, tapering or rounded at the base, 3–6 cm wide, coarsely serrate, hispid on both surfaces; heads 1 to few, in corymbs, on long peduncles, radiate; involucre hemispheric; phyllaries in 2 series, the outer somewhat foliaceous, the inner linear to linear-lanceolate, hispid to hirsute; receptacle usually hemispheric, the paleae 4–5 mm long, reaching the top of the corolla tube, ciliate near

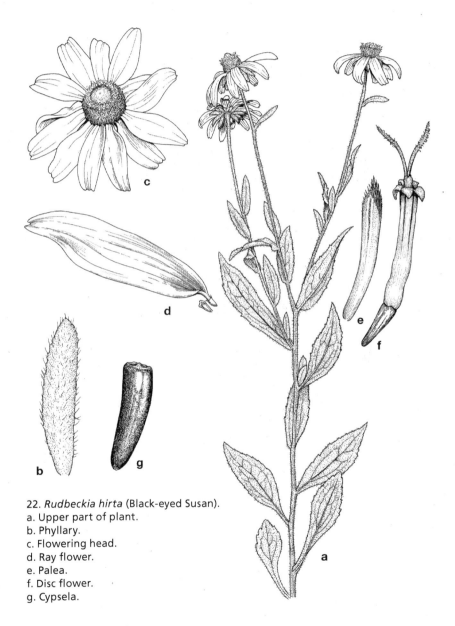

22. *Rudbeckia hirta* (Black-eyed Susan).
a. Upper part of plant.
b. Phyllary.
c. Flowering head.
d. Ray flower.
e. Palea.
f. Disc flower.
g. Cypsela.

the apex; ray flowers up to about 15, 2–4 cm long, yellow-orange, neutral; styles widely spreading; disc flowers numerous, 3–4 mm long, brown-purple, perfect, fertile, the corolla 5-lobed; cypselae 1.5–2.5 mm long, glabrous or nearly so; pappus absent.

Common Name: Black-eyed Susan.
Habitat: Prairies, pastures, black oak savannas, old fields.
Range: Maine to Illinois, south to Arkansas, Louisiana, and Georgia.
Illinois Distribution: Occasional in the eastern half of the state.

Although many botanists today believe *R. hirta* and *R. serotina* to be the same species, Nuttall's *R. serotina* is nothing at all like Linnaeus's *R. hirta*. Not only are there obvious differences of both the basal and cauline leaves of the two species, but the paleae and styles are significantly different.

Rays that are about 4 cm long with a dark, purplish base are var. *pulcherrima*. This species flowers from late June to early October.

6. **Rudbeckia grandiflora** (Sweet) F. Gmel. ex DC. Prodr. 5:556. 1836. Fig. 23.
Centrocarpha grandiflora Sweet, Brit. Flower Gard. Series 2, 1:plate 87. 1831.

Perennial herbs from fibrous roots and a woody caudex; stems erect, usually branched, to 1 m tall, with spreading pubescence; basal and lower cauline leaves ovate, acute at the apex, rounded or tapering to the base, to 35 cm long, to 10 cm wide, entire or sparingly serrate, gland-dotted; cauline leaves alternate, elliptic to lanceolate, acute at the apex, tapering or somewhat rounded at the short-petiolate or sessile base, to 25 cm long, to 8 cm wide, entire or sparingly serrate, hirsute, harshly scabrous, gland-dotted; head solitary, radiate, to 14 cm across, on a long peduncle; involucre hemispheric, 1–2 cm high, strigose; phyllaries in 2 series, 10–15 mm long, acuminate, strigose, gland-dotted; receptacle hemispheric to ovoid, the paleae 5–7 mm long, puberulent, awn-tipped; ray flowers 12–25, 3–7 cm long, 5–15 mm wide, reflexed, yellow, puberulent, gland-dotted, pistillate; disc flowers very numerous, purple, perfect, fertile, the corolla 3.5–5.0 mm long, with 5 reflexed lobes; cypselae flat, 2–3 mm long; pappus a crown of scales up to 0.5 mm long.

Common Name: Large-flowered coneflower.
Habitat: Adventive in a degraded prairie and along a railroad (in Illinois).
Range: Native to the southwestern United States; adventive in a few eastern states.
Illinois Distribution: Known only from DuPage County.

This handsome species with a large solitary flowering head has harshly scabrous leaves and strigose paleae.

This species flowers from August to October.

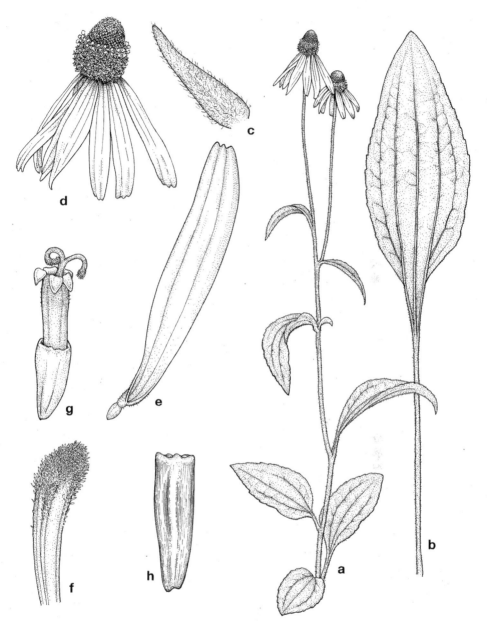

23. *Rudbeckia grandiflora* (Large-flowered coneflower).

a. Upper part of plant.
b. Basal leaf.
c. Phyllary.
d. Flowering head.
e. Ray flower.
f. Palea.
g. Disc flower.
h. Cypsela.

7. **Rudbeckia missouriensis** Engelm. ex Boynton & Beadle, Biltmore Bot. Stud. 1:17–18. 1901. Fig. 24.

Rudbeckia fulgida Ait. var. *missouriensis* (Engelm. ex Boynton & Beadle) Cronq. Rhodora 47:401. 1945.

Perennial herbs without rhizomes; stems erect, branched, to 75 cm tall, villous-hirsute; leaves basal and cauline, the basal ones linear to spatulate, acute at the apex, tapering to the base, to 20 cm long, to 2 cm wide, entire or serrulate, villous on both surfaces, the petioles to 15 cm long, the cauline ones linear to narrowly oblong, acute at the apex, tapering to the base, to 15 cm long, to 1.5 cm wide,

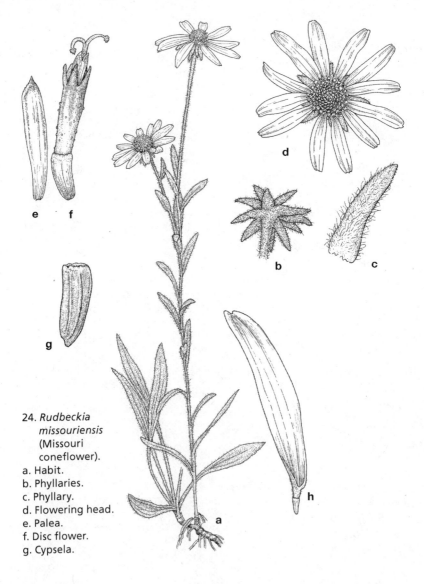

24. *Rudbeckia missouriensis* (Missouri coneflower).
a. Habit.
b. Phyllaries.
c. Phyllary.
d. Flowering head.
e. Palea.
f. Disc flower.
g. Cypsela.

entire, villous on both surfaces, sessile or short-petiolate; heads 1–10, in corymbs, to 4.5 cm across, radiate; involucre hemispheric; phyllaries up to 15, in 2 series, to 1.5 mm long, puberulent; receptacle hemispheric, the paleae 5.0–6.5 mm long, glabrous; ray flowers 10–15, 1.0–2.5 mm long, yellow, neutral; disc flowers numerous, perfect, fertile, purple-brown, the corolla 4.0–5.5 mm long, 5-lobed; cypselae 1.5–2.7 mm long, glabrous; pappus of minute scales about 0.1 mm long.

Common Name: Missouri coneflower.
Habitat: Hill prairies, glades.
Range: Illinois to Missouri, south to Texas, Louisiana, and Kentucky.
Illinois Distribution: Known only from Hardin, Monroe, and Randolph counties.

This species, a member of the *Rudbeckia fulgida* complex, differs by its linear cauline leaves that are entire.

Rudbeckia missouriensis lives on rocky slopes or in hill prairies. It flowers from July to October.

8. **Rudbeckia fulgida** Ait. Hort. Kew. 3:251. 1789. Fig. 25.

Perennial herbs from stolons, forming basal rosettes; stems erect, branched, to 1 m tall, villous-hirsute; leaves basal and cauline, the basal narrowly elliptic to oblanceolate, at least three times longer than wide, acute at the apex, tapering to the base, to 15 cm long, 2.0–4.5 cm wide, usually serrate, hirsute on both faces, on petioles to 10 cm long, the cauline alternate, lanceolate to lance-ovate, acute at the apex, tapering to the often sessile base, to 8 cm long, to 4 cm wide, more or less hirsute on both faces; heads 1 to few, in corymbs, to 3.5 cm across, radiate; involucre hemispheric; phyllaries up to 15, in 2 series, reflexed, 1.0–2.2 cm long, strigose or hispid; receptacle hemispheric, the paleae 4.5–6.0 mm long, ciliate along the margin; ray flowers 8–14, yellow, often with an orange base, 8–20 mm long, neutral; disc flowers numerous, perfect, bisexual, the corolla 3–4 mm long, brown-purple, 5-lobed; cypselae 2.5–3.5 mm long, glabrous or nearly so; pappus of minute scales 0.1–0.2 mm long.

Common Name: Orange coneflower.
Habitat: Woods.
Range: New York to Illinois, south to Alabama and Florida.
Illinois Distribution: Scattered in the southern half of Illinois.

The *Rudbeckia fulgida* complex in this work consists of *R. fulgida*, *R. missouriensis*, *R. tenax*, *R. sullivantii*, *R. palustris*, *R. umbrosa*, and *R. deamii*. Some botanists consider them all one variable species and some consider a few of them as varieties of *R. fulgida*. If one examines the phyllaries, length of rays, and leaf characteristics, each of these may be maintained as separate species, as I have done in this work.

Rudbeckia fulgida, *R. missouriensis*, and *R. tenax* are the only species in the complex in which the basal leaves are three times or more longer than wide. *Rudbeckia fulgida* differs from *R. missouriensis* by its toothed basal leaves, and from *R. tenax*

25. *Rudbeckia*
   fulgida (Orange
   coneflower).

a. Basal leaves.
b. Upper part of plant.
c. Phyllary.

d. Palea.
e. Disc flower.
f. Cypsela.

by its villous-hirsute stems, basal leaves more than 2 cm wide, and its phyllaries ciliate along the margin.

Rudbeckia fulgida flowers from July to October.

9. **Rudbeckia serotina** Nutt. Journ. Nat. Acad. Phila. 7:80–81. 1834. Fig. 26.

Perennial herbs from taproots; stems erect, branched, to 1 m tall, hispid; basal and cauline leaves present, the basal oblanceolate to lanceolate, acute at the apex, tapering or rounded at the base, 1–4 (–5) cm wide, entire or serrate, hispid on both surfaces, on petioles to 10 cm long, the cauline linear-lanceolate to oblanceolate, acute at the apex, tapering or rounded at the usually sessile base, up to 3 cm wide, entire or finely serrate, hispid on both surfaces; heads 1 to few, in corymbs, on short to long peduncles, radiate; involucre hemispheric; phyllaries in 2 series, linear to linear-lanceolate, hispid to hirsute; receptacle usually hemispheric, the paleae 4.5–6.0 mm long, not reaching the tip of the corolla tube, ciliate from the middle to the apex; ray flowers up to about 15, 1–5 cm long, yellow-orange, neutral; styles ascending; disc flowers numerous, 3–5 mm long, bisexual, fertile, brown-purple; cypselae 1.5–2.8 mm long, glabrous or nearly so; pappus absent.

This species, often considered the same as *R. hirta*, differs by its basal leaves oblanceolate to lanceolate and 1–4 cm wide, its cauline leaves that are linear-lanceolate to oblanceolate, finely serrate, and up to 3 cm wide, its paleae that are 4.5–6.0 mm long and ciliate from the middle to the apex and not reaching the top of the tube of the corolla, and its ascending styles.

Two varieties occur in Illinois:

a. Rays 1.0–3.5 cm long.................................9a. *R. serotina* var. *serotina*
a. Rays 3.5–5.0 mm long.............................9b. *R. serotina* var. *lanceolata*

9a. **Rudbeckia serotina** Nutt. var. **serotina**
Rays 1.0–3.5 cm long.

Common Name: Black-eyed Susan.
Habitat: Old fields, prairies, dry woods, pastures, disturbed soil.
Range: Nova Scotia to British Columbia, south to California, Texas, and Florida.
Illinois Distribution: Common throughout the state.

This variety flowers from June to October.

9b. **Rudbeckia serotina** Nutt. var. **lanceolata** (Bisch.) Fern. & Schub. Rhodora 50:175. 1948.
Corolla 3.5–5.0 cm long.

Common Name: Brown-eyed Susan.
Habitat: Cherty woods.
Range: Apparently native in Illinois and Missouri; adventive elsewhere.
Illinois Distribution: Known only from Union County.

26. *Rudbeckia
serotina* (Black-
eyed Susan).

a. Upper part of plant.
b. Lower part of plant.
c. Leaf.
d. Flowering head.

e. Phyllary.
f. Palea.
g. Disc flower.
h. Cypsela.

This variety differs from typical var. *serotina* by its longer rays. Because of the rays, it is a very beautiful plant when in flower. Fernald and Schubert believed this variety to be native to Missouri and Illinois. The only Illinois record is in a cherty woods beneath *Pinus echinata* in the Pine Hills of Union County.

Rudbeckia serotina var. *lanceolata* flowers from June to October.

10. **Rudbeckia tenax** Boynton & Beadle, Fl. S.E.U.S. 1257. 1903. Fig. 27.

Perennial herbs from stolons, sometimes forming basal rosettes; stems erect branched, to 75 cm tall, glabrous or sometimes strigose; leaves basal and cauline, the basal ovate-lanceolate to elliptic, acute at the apex, tapering to the base, to 10 cm long, 1–2 cm wide, 3 times longer than wide, serrate, strigose or sometimes hispid on both surfaces, the petioles to 6 cm long, the cauline ovate-lanceolate to lanceolate to elliptic-lanceolate, to 8 cm long, 1–3 cm wide, acute to acuminate at the apex, tapering to the base, the lower ones on winged petioles, the uppermost sessile; heads 1 to few, in corymbs, to 3 cm across, radiate; involucre hemispheric;

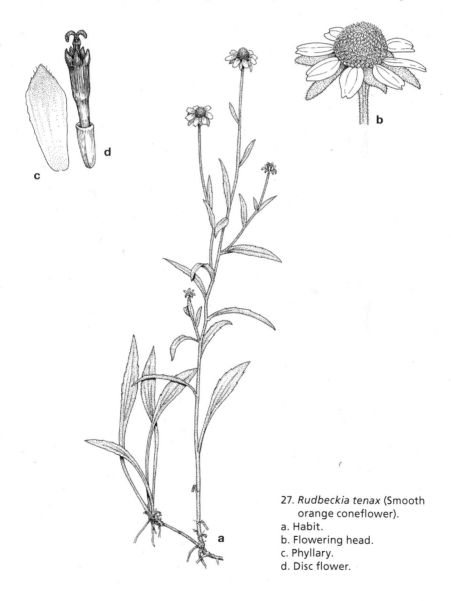

27. *Rudbeckia tenax* (Smooth orange coneflower).
a. Habit.
b. Flowering head.
c. Phyllary.
d. Disc flower.

phyllaries up to 15, in 2 series, reflexed, 1–2 cm long, densely hispid; receptacle hemispheric, the paleae 4.0–5.5 mm long, eciliate; ray flowers 6–14, 8–20 mm long, yellow or orange, neutral; disc flowers numerous, bisexual, perfect, the corolla 3–4 mm long, brown-purple; cypselae 2.5–3.5 mm long, glabrous or nearly so; pappus of minute scales 0.1–0.2 mm long.

Common Name: Smooth orange coneflower.
Habitat: Wet ground.
Range: Indiana and Illinois, south to Alabama.
Illinois Distribution: Confined to the southern one-sixth of the state.

This species is in the complex that includes *R. fulgida* and *R. missouriensis* where the basal leaves are about three times longer than broad and the rays are usually between 5 and 20 mm long.

   Rudbeckia tenax differs by its serrate basal leaves 1–2 cm wide and its paleae, which are eciliate.

   Rudbeckia tenax flowers from July to October.

   11. **Rudbeckia sullivantii** Boynton & Beadle, Biltmore Bot. Studies 1:15–16. 1901. Fig. 28.
Rudbeckia speciosa Wend. var. *sullivantii* (Boynton & Beadle) B.L. Robins. Rhodora 10:68. 1908.
Rudbeckia fulgida Ait. var. *sullivantii* (Boynton & Beadle) Cronq. Rhodora 47:401. 1945.

   Perennial herbs from stolons, forming basal rosettes; stems erect, branched, to 1 m tall, hirsute with spreading hairs; leaves basal and cauline, the basal broadly elliptic to ovate, acute at the apex, rounded at the base, about twice as long as wide, to 15 cm long, 1.5–3.0 cm long, sharply serrate, glabrous or sparsely pubescent on both surfaces, on petioles to 10 cm long, the cauline lanceolate to ovate, acute at the apex, tapering or rounded at the base, the upper much smaller than the lower, entire or serrate, glabrous or sparsely pubescent on both surfaces, the largest to 8 cm long, to 4 cm wide, petiolate; heads 1 to few, in corymbs, to 3.5 cm across, radiate; involucre hemispheric; phyllaries up to 15, in 2 series, spreading to reflexed, 1–2 cm long, glabrous or sparsely pubescent; receptacle hemispheric, the paleae 5–6 mm long, glabrous, eciliate; ray flowers 10–20, 25–40 mm long, yellow or orange, neutral; disc flowers numerous, bisexual, perfect, brown-purple, the corolla 3.5–4.5 mm long; cypselae 2.5–3.5 mm long, glabrous or nearly so; pappus of a short crown.

Common Name: Sullivant's coneflower.
Habitat: Wet woods, wooded swamps, calcareous fens.
Range: New York to Michigan, south to Arkansas and West Virginia.
Illinois Distribution: Scattered in Illinois, but not common.

28. *Rudbeckia sullivantii* (Sullivant's coneflower).
a. Habit.
b. Phyllary.
c. Cluster of phyllaries.
d. Flowering head.
e. Ray flower.
f. Disc flower with palea.
g. Cypsela.

This species and *R. palustris* differ from other species in the *R. fulgida* complex by having their basal leaves less than three times longer than wide and with the upper cauline leaves much smaller than the lower cauline leaves. *Rudbeckia sullivantii* differs from *R. palustris* by its longer rays and eciliate paleae. Early Illinois botanists called this plant *R. speciosa*.

This is a species of swamps and calcareous fens.

Rudbeckia sullivantii flowers from July to October.

12. **Rudbeckia palustris** Eggert ex Boynton & Beadle, Biltmore Bot. Studies 1:16–17. 1901. Fig. 29.

Rudbeckia fulgida Ait. var. *palustris* (Eggert ex Boynton & Beadle) Perdue, Rhodora 59:297–298. 1957.

Perennial herbs from stolons, forming basal rosettes; stems erect, branched, to 1.2 m tall, glabrous or with spreading hairs; leaves basal and cauline, the basal narrowly elliptic to ovate, about twice as long as wide, acute at the apex,

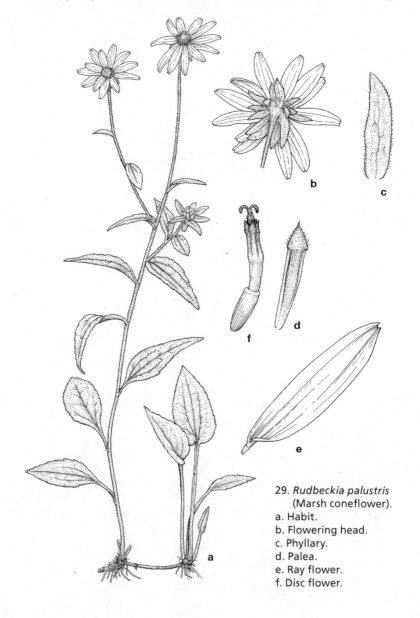

29. *Rudbeckia palustris*
   (Marsh coneflower).
a. Habit.
b. Flowering head.
c. Phyllary.
d. Palea.
e. Ray flower.
f. Disc flower.

tapering to the base, to 15 cm long, 2–4 cm wide, entire or serrate, glabrous or spreading pubescent on both surfaces, the petioles to 10 cm long, the cauline lanceolate to ovate, acute at the apex, tapering or somewhat auriculate at the base, the upper much smaller than the lower and usually sessile or the lower petiolate, entire or serrate, sparsely pubescent on both surfaces, the longest to 8 cm long, to 3.5 cm wide; heads 1 to few, to 3.5 cm wide, in corymbs, radiate; involucre hemispheric; phyllaries up to 15, in 2 series, reflexed, 0.8–2.0 cm long, glabrous on the inner face, pubescent on the outer face; receptacle hemispheric, the paleae 5–6 mm long, glabrous, ciliate; ray flowers 8–15, 15–25 mm long, yellow or orange, neutral; disc flowers numerous, bisexual, perfect, brown-purple, the corolla 3.5–4.5 mm long; cypselae 2.5–3.5 mm long, glabrous or nearly so; pappus a short crown.

Common Name: Marsh coneflower.
Habitat: Wet woods, wet fields
Range: Illinois to Missouri, south to Texas and Arkansas.
Illinois Distribution: Known from Jackson and Union counties.

This species is similar to *R. sullivantii*, but differs by its shorter rays and its ciliate paleae.

    Rudbeckia palustris flowers from July to September.

    13. **Rudbeckia umbrosa** Boynton & Beadle, Biltmore Bot. Studies 1:16. 1901. Fig. 30.
Rudbeckia fulgida Ait. var. *umbrosa* (Boynton & Beadle) Cronq. Rhodora 47:400. 1945.

Perennial herbs from thick rhizomes, forming basal rosettes; stems erect, branched, to 1 m tall, glabrous or sparsely pubescent with spreading hairs; leaves basal and cauline, the basal ovate, up to twice as long as wide, acute at the apex, rounded to cordate at the base, to 15 cm long, 3.5–5.0 cm wide, coarsely dentate, glabrous or sparsely pubescent on both surfaces, the petioles to 10 cm long, the cauline leaves lanceolate to ovate, acute at the apex, tapering or rounded at the base, the upper gradually smaller than the lower and usually sessile, the lower petiolate, entire or coarsely serrate, pubescent on both surfaces, the largest to 9 cm long, to 4 cm wide; heads 1 to few, to 3.5 cm across, in corymbs, radiate; involucre hemispheric; phyllaries up to 15, in 2 series, reflexed, 1.0–2.2 cm long, glabrous or sparsely pubescent; receptacle hemispheric to conic, the paleae 4.5–5.5 mm long, glabrous, ciliate; ray flowers 8–12, 10–30 mm long, yellow or orange, neutral; disc flowers numerous, bisexual, perfect, purple, the corolla 3.0–4.5 mm long; cypselae 3–4 mm long, glabrous or nearly so; pappus a prominent crown.

Common Name: Shady coneflower.
Habitat: Bottomland forests.

30. *Rudbeckia umbrosa* (Shady
   coneflower).
a. Habit.
b. Flowering head.
c. Phyllary.
d. Palea.
e. Disc flower.

Range: Virginia to Missouri, south to Arkansas, Mississippi, and Georgia.
Illinois Distribution: Mostly along the Mississippi and Ohio rivers; not common.

The distinguishing features of this species are its coarsely dentate basal leaves up to twice as long as wide, its upper cauline leaves not noticeably smaller than the lower cauline leaves, and the relatively few rays per flowering head.

   Rudbeckia umbrosa flowers from August to October.

14. **Rudbeckia deamii** Blake, Rhodora 19:113–115. 1917. Fig. 31.
Rudbeckia fulgida Ait. var. *deamii* (Blake) Perdue, Rhodora 59:297. 1959.

Perennial herbs from stolons, forming basal rosettes; stems erect, branched, to
1 m tall, densely villous-hirsute, the hairs pointing downward; leaves basal and
cauline, the basal oval to ovate, up to twice as long as wide, acute at the apex,

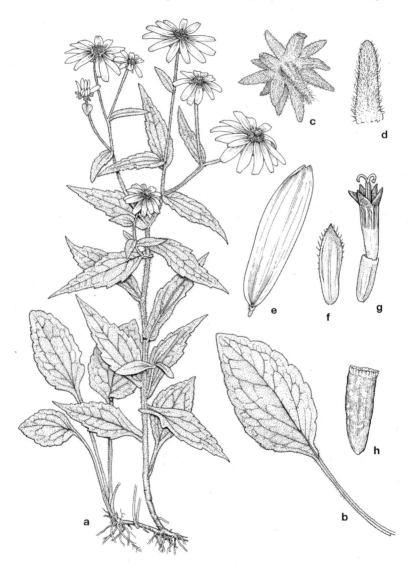

31. *Rudbeckia*
    deamii (Deam's
    coneflower).

a. Habit.
b. Lower leaf.
c. Phyllaries.
d. Phyllary.

e. Ray flower.
f. Palea.
g. Disc flower.
h. Cypsela.

tapering or rounded at the base, to 8 cm long, to 3.5 cm wide, coarsely crenate, pubescent on both surfaces, the petioles up to 8 cm long, the cauline leaves elliptic to broadly lanceolate, acute at the apex, auriculate at the base, the upper gradually smaller than the lower and usually sessile, the lower sparsely but sharply serrate, densely villous-hirsute on both surfaces, the largest to 10 cm long, to 4 cm wide; heads 1 to few, up to 3 cm across, in corymbs, radiate; involucre hemispheric; phyllaries up to 15, in 2 series, reflexed, 1–2 cm long, densely pubescent on both surfaces; receptacle hemispheric, the paleae 4–5 mm long, glabrous, ciliate; ray flowers 12–20, 15–25 mm long, yellow or orange, neutral; disc flowers numerous, brown-purple, bisexual, perfect, the corolla 3.0–4.5 mm long; cypselae 3–4 mm long, glabrous or nearly so; pappus of minute scales.

Common Name: Deam's coneflower.
Habitat: Along streams (in Illinois).
Range: Ohio, Indiana, and Illinois.
Illinois Distribution: Rare in the southeastern counties.

This is one of the more distinct species because of its densely retrorse-hirsute stems, its coarsely crenate basal leaves, and its clasping cauline leaves. The phyllaries are also densely pubescent on both surfaces.

This species has the most restricted range in the *R. fulgida* complex. It flowers during August and September.

15. **Rudbeckia speciosa** Wend. Ind. Sem. Hort. Marburg. 1828. Fig. 32.
Rudbeckia hirta L. var. *speciosa* (Wend.) Perdue, Rhodora 59:297. 1957.

Perennial herbs from slender rhizomes and stolons, forming basal rosettes; stems erect, branched, to 1 m tall, glabrous or sparsely hirsute with spreading hairs; leaves basal and cauline, the basal broadly lanceolate to ovate, up to twice as long as wide, acute at the apex, rounded at the base, up to 12 cm long, up to 6.5 cm wide, entire or finely crenate, pubescent on both surfaces, the petioles to 15 cm long, the cauline leaves elliptic to lanceolate, acute at the apex, tapering to the base or subauriculate, coarsely serrate, glabrous or scabrous on either surface, to 10 cm long, to 3.5 cm wide, the upper gradually reduced in size from the lower; heads 1 to few, to 3.5 cm long, in corymbs, radiate; involucre hemispheric; phyllaries up to 15, in 2 series, reflexed, 1–2 cm long, glabrous or sparsely pubescent; receptacle hemispheric, the paleae 4–5 mm long, glabrous, usually ciliate; ray flowers 12–20, 20–40 mm long, yellow or orange, neutral; disc flowers numerous, brown-purple, bisexual, perfect, the corolla 3.5–4.5 mm long; cypselae 3.5–5.0 mm long, glabrous or nearly so; pappus a short crown.

Common Name: Showy coneflower.
Habitat: Moist woods, prairies, sand prairies.
Range: Ontario to Wisconsin, south to Arkansas and Georgia.
Illinois Distribution: Occasional throughout the state.

32. *Rudbeckia speciosa* (Showy coneflower).
a. Upper part of plant.

b. Lower part of plant showing basal leaves.
c. Outer phyllary.

d. Inner phyllary.
e. Flowering head.
f. Palea.
g. Disc flower.

The species and *R. sullivantii* have the largest and showiest rays of any species in the complex. It differs from *R. sullivantii* by its upper cauline leaves just gradually reduced in size from the lower cauline leaves and by its usually ciliate paleae.

This species flowers from August to October.

40. **Ratibida** Raf.—Prairie Coneflower

Perennial herbs; stems erect, branched, gland-dotted; leaves basal and cauline, petiolate, pinnately lobed to pinnatifid, usually gland-dotted; heads solitary or sometimes several in corymbs, radiate, on long peduncles; involucre rotate; phyllaries in 2–3 series, unequal, somewhat reflexed, persistent; receptacle columnar, paleate, strigose, gland-dotted; ray flowers up to 20, yellow or yellow and maroon, neutral; disc flowers numerous, bisexual, fertile, tubular, yellow-green to purplish, the corolla 5-lobed; cypselae flat, sometimes winged along the ciliate or laciniate margins, glabrous or pubescent; pappus of 1–2 short teeth.

Ratibida consists of seven species, all in North America.

This genus differs from *Rudbeckia* and *Dracopis,* the other yellow-flowered coneflowers in Illinois, by its flat, slightly winged cypselae. *Rudbeckia* has a 4-winged cypsela, and *Dracopis* has a terete cypsela. In addition, the pappus is absent in *Dracopis.*

Two species occur in Illinois:

1. Divisions of the deeply pinnatifid cauline leaves linear; rays 20–35 mm long; disc 1–5 cm high . . . . .1. *R. columnifera*
1. Divisions of the pinnatifid cauline leaves lanceolate; rays 15–60 mm long; disc 1.0–2.5 cm high . . . . .2. *R. pinnata*

1. **Ratibida columnifera** (Nutt.) Wooton & Standl. Contr. U.S. Natl. Herb. 19:706. 1915. Fig. 33.

Rudbeckia columnifera Nutt. Cat. Pl. Upper Louisiana, number 75. 1813.

Rudbeckia columnaris Sims, Bot. Mag. pl. 1601. 1813.

Obelisaria pulcherrima DC. Prodr. 5:559. 1836.

Ratibida columnaris (Sims) D. Don in Sweet, Brit. Fl. Gard. 2:361. 1838.

Lepachys columnaris (Sims) Torr. & Gray, Fl. N. Am. 2:315. 1842.

Lepachys columnaris (Sims) Torr. & Gray var. *pulcherrima* Torr. & Gray, Fl. N. Am. 2:315. 1842.

Lepachys columnifera (Nutt.) J.F. Macbr. Contr. Gray Herb. 65:45. 1922.

Lepachys columnifera (Nutt.) J.F. Macbr. var. *pulcherrima* (DC.) Rydb. Fl. Plains N. Am. 838. 1932.

Ratibida columnifera Nutt. f. *pulcherrima* (DC.) Fern. Rhodora 40:353. 1938.

Perennial herb from a taproot; stems erect, branched, to 1 m tall, strigose, scabrous; leaves alternate, deeply 1- to 2-pinnatifid, to 15 cm long, with up to 11 (–13) lobes, the lobes linear to narrowly oblong, entire or cleft, hirsute on both surfaces, sessile or on short petioles; heads radiate, solitary or up to 15 in a corymb, on long peduncles elevated well above the leaves; involucre rotate, 8–15 mm across; phyllaries in 2–3 series, unequal, the outer linear, up to 14 mm long, the inner narrowly ovate, to 3 mm long; receptacle columnar, paleate, the paleae 2.5–3.5 mm long, resinous, canescent; ray flowers 4–10 (–12), up to 4 cm long, reflexed, yellow, often maroon at base, rarely all purple, neutral; disc columnar, up to 3 cm long, with numerous tubular flowers greenish to purple, up to 2.5 mm high, 5-lobed, perfect, fertile; cypselae oblong, flat, narrowly winged, glabrous, or sometimes ciliate, 1.5–3.0 mm long; pappus of 1–2 subulate teeth, or rarely the teeth absent.

Common Name: Prairie coneflower; Mexican hat; long-headed coneflower.
Habitat: Mostly along railroads (in Illinois).
Range: Ontario to British Columbia, south to Montana, Arizona, Louisiana, and
    North Carolina; adventive in Illinois.
Illinois Distribution: Scattered in the northern two-thirds of the state.

This species differs from *R. pinnata* by its long, columnar disc and the presence of a
taproot rather than a rhizome.

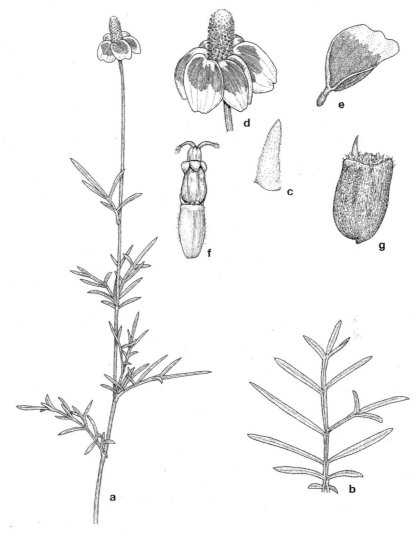

33. *Ratibida columnifera*
    (Prairie coneflower).
a. Upper part of plant.

b. Leaf.
c. Phyllary.
d. Flowering head.

e. Ray flower.
f. Disc flower.
g. Cypsela.

Plants with rays all purple have been called f. *pulcherrima*. This form has been found in Illinois.

Some botanists have called this species *R. columnaris*. The basionym for *R. columnifera* by Nuttall and for *R. columnaris* by Sims were both published in 1813, but apparently Nuttall's epithet was earlier in the year.

Ratibida columnifera is a common plant of the prairies and Great Plains, but is adventive in states to the east, including Illinois. Because of its pretty flowering heads, it is sometimes used as a garden ornamental, and its seeds are often in seed mixes for prairie restorations.

This species flowers from June to November.

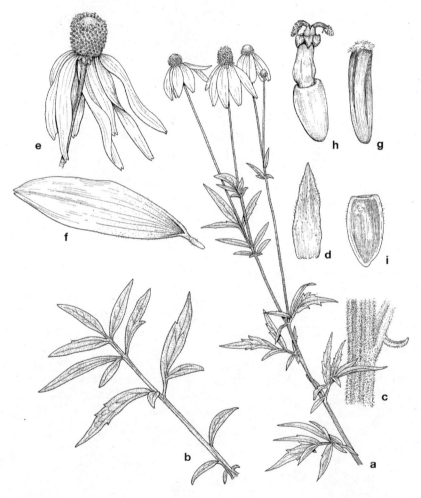

34. *Ratibida pinnata*
    (Drooping
    coneflower).
a. Upper part of plant.

b. Leaf.
c. Portion of stem.
d. Phyllary.
e. Flowering head.

f. Ray flower.
g. Palea.
h. Disc flower.
i. Cypsela.

2. **Ratibida pinnata** (Vent.) Barnh. Bull. Torrey Club 24:410. 1897. Fig. 34.
Rudbeckia pinnata Vent. Descr. Pl, Nouv. Pl. 71. 1802.
Lepachys pinnata (Vent.) Torr. & Gray, Fl. N. Am. 2:314. 1842.

Perennial herbs from rhizomes; stems erect, branched or unbranched, to 1.5 m tall, strigose, scabrous; leaves alternate, to 40 cm long, pinnate or pinnatifid, with 3, 5, 7, or 9 lobes, the lobes lanceolate to ovate, entire, dentate, of cleft, strigose on both surfaces, glandular-dotted, petiolate or the uppermost sessile; heads 1–12, radiate, elevated well above the leaves on strigose peduncles; involucre rotate, 8–12 mm across; phyllaries in 2–3 series, unequal, the outer linear, reflexed, to 15 mm long, the inner narrowly ovate, reflexed, to 6 mm long; receptacle short-columnar, paleate, the paleae to 5 mm long, resinous; ray flowers up to 15, yellow, to 6 cm long, notched at the tip, strongly reflexed, neutral; disc globose to ovoid, with numerous greenish yellow or purplish tubular flowers, 2.5–4.0 mm long, 5-lobed, bisexual, perfect; cypselae narrowly lanceoloid, 2–4 mm long, glabrous, sometimes ciliate; pappus absent or reduced to 1 or 2 minute teeth.

Common Name: Drooping coneflower; yellow coneflower; gray coneflower.
Habitat: Prairies, rocky glades.
Range: New Jersey and Ontario to South Dakota, south to Oklahoma, Louisiana, and Florida.
Illinois Distribution: Common in the northern three-fourths of Illinois, occasional elsewhere.

This species is distinguished by its pinnate or pinnatifid leaves, its strongly reflexed yellow rays, and its globose to ovoid disc.

Ratribida pinnata flowers from July to October.

Subtribe Ecliptinae Less.

Annuals, biennials, perennials, and shrubs; leaves basal and/or cauline, opposite or alternate, often lobed; heads radiate or less commonly discoid, borne singly or into a variety of inflorescences; involucre cylindric, campanulate, hemispheric, or rotate; phyllaries usually in 2–4 series, subequal or unequal, the outer usually shorter than the inner; receptacle convex, conic, or flat, paleate; ray flowers up to 40, rarely absent, usually pistillate and fertile, variously colored; disc flowers 4 to numerous, perfect, fertile or sometimes only staminate, usually 5-lobed, variously colored; cypselae obpyramidal to prismatic to terete to ovoid, sometimes flat or nearly so, sometimes ribbed, sometimes winged; pappus of scales or bristles, rarely absent.

This subtribe consists of 75 genera and around 600 species, with 31 genera and 124 species in the United States.

Key to the Genera of Subtribe Ecliptinae in Illinois.
1. Heads discoid; leaves opposite . . . . .48. *Melanthera*
1. Heads radiate; leaves opposite, alternate, and/or basal.

2. Rays yellow or orange.
   3. Leaves mostly opposite.
      4. Pappus of numerous capillary bristles; plants creeping ...... 51. *Calyptocarpus*
      4. Pappus of awns, scales, 1–3 short bristles, or absent; plants erect (procumbent in *Sanvitalia*).
         5. Disc flowers sterile with poorly developed cypselae .... 43. *Silphium*
         5. Disc flowers fertile with well-developed cypselae.
            6. Ray flowers persistent on the cypselae.................... 41. *Heliopsis*
            6. Ray flowers deciduous from the cypselae.
               7. Pappus of 1–3 short bristles, or absent .... 50. *Acmella*
               7. Pappus of awns.
                  8. Cypselae 3-angled; pappus of cypselae of ray flowers with 3 awns... 42. *Sanvitalia*
                  8. Cypselae 2- or 4-angled; pappus of cypselae of ray flowers with 2 or 4 awns.................................... 45. *Verbesina*
   3. Leaves mostly alternate.
      9. Pappus of the cypselae of the disc flowers with short teeth .... 43. *Silphium*
      9. Pappus of the cypselae of the disc flowers with 2–8 awns.
         10. Disc flowers sterile; phyllaries in 2–3 series; stems unwinged ... 43. *Silphium*
         10. Disc flowers fertile; phyllaries in 1–3 series; stems winged or unwinged.
            11. Annuals; stems unwinged; leaves with a basal flange of tissue; flowering heads 1–3; phyllaries 12–18, spreading; rays 12–15, 3-notched at the tip; disc flowers 80–150; cypselae of disc flowers winged, of ray flowers wingless................................... 47. *Ximenesia*
            11. Perennials; stems 4-winged; leaves without a basal flange of tissue; flowering heads up to 50; phyllaries 8–12, reflexed or erect; rays 2–13, not 3-notched at the tip; disc flowers 40–100; cypselae of all flowers broadly winged .... 46. *Actinomeris*
2. Rays neither yellow nor orange.
   12. Leaves opposite; rays white.
      13. Outer phyllaries stipitate-glandular; plants usually at least 1.5 m tall .... 45. *Verbesina*
      13. Outer phyllaries eglandular; plants usually less than 1.5 m tall. ....48. *Eclipta*
   12. Leaves alternate; rays white or pink or rose or purple.
      14. Rays white .... 45. *Verbesina*
      14. Rays pink, rose, or purple................................. 44. *Echinacea*

41. Heliopsis Pers.—Oxeye Sunflower

Perennial herbs; stems usually erect; leaves opposite, simple, cauline, 3-nerved from the base, petiolate; heads radiate, borne singly both axillary and terminal; involucre hemispheric or campanulate or turbinate; phyllaries in 2–3 series, the outer foliaceous and spreading, the inner narrower and erect; receptacle conical or convex, paleate, more or less persistent and often chartaceous; ray flowers up to 20, yellow or orange, pistillate, fertile, becoming chartaceous with age and persistent on the cypselae; disc flowers numerous, perfect, yellow to purplish, the corolla 5-lobed; cypselae 3- or 4-angled or more or less terete; pappus of 2–4 minute teeth, or absent.

About 18 species in North and South America comprise this genus. Only the following occurs in Illinois.

1. **Heliopsis helianthoides** (L.) Sweet, Hort. Bot. 487. 1827. Figs. 35a, 35b.
Buphthalmum helianthoides L. Sp. Pl. 2:904. 1753.

Perennial herb from rhizomes; stems erect, branched, to 1.5 m tall, glabrous or
harshly scabrous; leaves opposite, ovate-lanceolate to ovate, acute to acuminate at
the apex, narrowed to the base or sometimes rounded or subcordate, to 12 cm long,
to 8 cm wide, glabrous or strongly scabrous, coarsely serrate or dentate, petiolate;
heads radiate, 1–15, axillary or terminal, on glabrous or scabrous peduncles, up
to 2.5 cm long; involucre hemispheric; phyllaries in 2–3 series, the outer up to 6
mm wide, obtuse at the apex, glabrous or pubescent; receptacle paleate, the paleae
obtuse at the apex; ray flowers up to 18, yellow or orange, 2–4 cm long, pistillate,
fertile, becoming chartaceous with age; disc flowers up to 75, yellowish, perfect, the
corolla 5-lobed, 4–5 mm long, the disc up to 2.5 cm across; cypselae 3- or 4-angled,
glabrous or pubescent, 4–5 mm long; pappus of 2–4 minute teeth, or absent.

Two varieties, which often intergrade, occur in Illinois: .
a. Leaves glabrous or scarcely scabrous on the upper surface........................
.... 1a. *H. helianthoides* var. *helianthoides*
a. Leaves harshly scabrous on the upper surface......... 1b. *H. helianthoides* var. *scabra*

1a. **Heliopsis helianthoides** (L.) Sweet var. **helianthoides** Fig. 35a.
Heliopsis laevis Pers. Syn. 2:473. 1807.

Leaves glabrous or scarcely scabrous on the upper surface, to 8 cm wide; cypselae glabrous; phyllaries usually glabrous; pappus usually absent.

Common Name: Oxeye sunflower; false sunflower.
Habitat: Open woods, prairies, mesic savannas, prairie fens.
Range: Massachusetts to Wisconsin, south to Missouri, Louisiana, and Georgia.
Illinois Distribution: Common throughout the state.

This common variety resembles species of *Helianthus*, but differs by its ray flowers
that become chartaceous and persist on the cypselae and by its shorter and often
more obtuse phyllaries.

Heliopsis helianthoides is sometimes planted as an ornamental. It flowers from
June to October.

1b. **Heliopsis helianthoides** (L.) Sweet var. **scabra** (Dunal) Fern. Rhodora
44:340. 1942. Fig. 35b.
Heliopsis scabra Dunal, Mem. Mus. Paris 5:56. 1819.
Heliopsis laevis Pers. var. *scabra* (Dunal) Torr. & Gray, Fl. N. Am. 2:303. 1842.

Leaves lance-ovate, harshly scabrous on the upper surface, to 5 cm wide; cypselae pubescent when young, at least on the angles; phyllaries canescent; pappus
usually of 2–4 minute teeth.

Plants with extremely harshly scabrous upper leaf surfaces also usually differ
from var. *helianthoides* by having narrower leaves, canescent phyllaries, and a

35a. *Heliopsis*
     helianthoides
     var. *helianthoides*
     (Oxeye sunflower).

a. Upper part of plant.
b. Phyllaries.
c. Phyllary.
d. Flowering head.

e. Ray flower.
f. Palea.
g. Disc flower.
h. Cypsela.

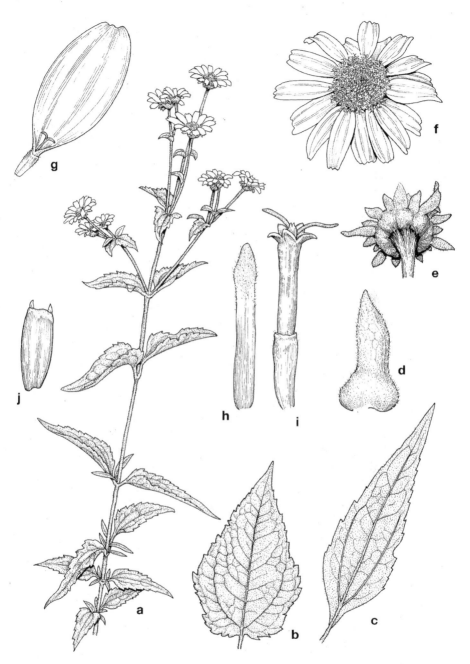

35b. *Heliopsis*
helianthoides
var. *scabra*
(Rough orange
sunflower).

a. Habit.
b, c. Leaf.
d. Phyllary.
e. Phyllaries.
f. Flowering head.

g. Ray flower.
h. Palea.
i. Disc flower.
j. Cypsela.

pappus consisting of 2–4 minute teeth. Dunal felt it was distinct enough to merit recognition as a distinct species. There is intergradation throughout the state between the two varieties.

Variety *scabra* flowers from June to October.

42. Sanvitalia Lam.—Creeping Zinnia

Annual or perennial herbs; stems procumbent to erect, branched; leaves cauline, simple, opposite, petiolate or sessile, pubescent; head radiate, borne singly; involucre hemispheric to rotate; phyllaries in 2–3 series, the outer foliaceous; receptacle convex to conic, paleate; ray flowers up to 20, usually yellow, becoming chartaceous, pistillate, fertile; disc flowers up to 60, tubular, yellow or orange, perfect, fertile, the corolla 5-lobed; cypselae terete to obscurely angled, usually somewhat tuberculate; pappus of 3–4 awns, persistent on the cypselae.

Five species are in the genus. Two of them are native to the southwestern United States, the others to Mexico, Central America, and South America. One is an adventive garden escape in the United States.

I. **Sanvitalia procumbens** Lam. Journ. Hist. Nat. 2:176. 1792. Fig. 36.

Annual herb from a taproot; stems prostrate to ascending to erect, branched from the base, more or less pubescent, to 15 cm tall; leaves cauline, simple, opposite, lance-linear to ovate, acute at the apex, tapering to the base, mostly entire or occasionally serrulate, pubescent on both surfaces, to 6 cm long, to 3 cm wide; involucre hemispheric, to 1.2 cm across; phyllaries in 2–3 series, the outer foliaceous, linear to lanceolate, the inner subulate; receptacle convex, paleate; head solitary, radiate, on a short peduncle; ray flowers up to 20, to 1 cm long, yellow, pistillate, fertile; disc flowers up to 60, yellow, tubular, perfect, the corolla 5-lobed; cypselae of ray flowers 3-angled, 2- to 3-nerved, usually glabrous, of outer disc flowers 4-angled, wingless, 2–3 mm long, glabrous or nearly so, of inner disc flowers flattened, 1- or 2-winged, 2–3 mm long, glabrous or nearly so; pappus of ray flowers with 3–4 awns 1–3 mm long, those of the outer disc flowers with 3–4 mm awns less than 1 mm long; those of the inner disc flowers flat, 1- or 2-winged, with a pappus of 3–4 awns less than 1 mm long.

Common Name: Creeping zinnia.
Habitat: Disturbed soil.
Range: Native to Mexico and Central America; adventive in a few states.
Illinois Distribution: Escape from gardens in Cook and Jackson counties.

This handsome small plant is a popular ornamental that sometimes escapes from cultivation but rarely persists. The small colony in Jackson County has escaped to the edge of a woods where it has maintained itself for several years.

The cypselae are of three kinds: those of the ray flowers are trigonous and 2- or 3-nerved, with a pappus of 3–4 awns 1–3 mm long; those of the outer disc flowers are 4-angled, wingless, with a pappus of 3–4 awns less than 1 mm long; those of

36. *Sanvitalia procumbens* (Creeping zinnia).
a. Upper part of plant.
b. Phyllary.
c. Side view of flowering head.
d. Face view of flowering head.
e. Ray flower.
f. Disc flower with palea.
g. Cypsela.

the inner disc flowers flat, 1- or 2-winged, with a pappus of 3–4 awns less than 1 mm long.

Sanvitlia procumbens flowers during July.

43. Silphium L.—Rosinweed

Perennial herbs from rhizomes and taproots, usually resinous; stems erect, branched, terete or 4-sided; leaves basal and/or cauline, simple, usually cauline, often opposite, less commonly whorled or alternate, sometimes basal; heads 1 to several, radiate, borne in corymbs or panicles; involucre hemispheric to campanulate; phyllaries in 2–4 series, the outer foliaceous, the inner smaller; receptacle flat, paleate; ray flowers up to 40, in 2–4 series, yellow, pistillate, fertile; disc flowers numerous, yellow, pistillate, tubular, the corolla 5-lobed; cypselae flattened, usually winged, notched at the apex; pappus absent or of 2 very short awns.

Fifteen species, all North American, are in the genus, several of them in Illinois.

1. Leaves connate-perfoliate; stems conspicuously 4-angled . . . . 4. *S. perfoliatum*
1. Leaves not connate-perfoliate; stems terete or only slightly angular.
  2. Leaves deeply pinnatifid.
    3. Phyllaries appressed, obtuse to acute at the apex . . . . 3. *S. pinnatifidum* ·
    3. Phyllaries often reflexed, acuminate to caudate at the apex . . . . 1. *S. laciniatum*
  2. Leaves entire or serrate.
    4. Leaves all basal . . . . 2. *S. terebinthinaceum*
    4. Leaves cauline (basal leaves withered at anthesis, except *S. trifoliatum*.)
      5. Stems, leaves, and phyllaries glabrous or nearly so; stems glaucous . . . .
      . . . . 6. *S. speciosum*
      5. Stems, leaves, and phyllaries hispid, or at least scabrous; stems not glaucous.
        6. Cauline leaves opposite; basal leaves withered at anthesis.
          7. Disc flowers up to 130 (–140) per head; ray flowers up to 20 (–22) per
          head . . . . 5. *S. integrifolium*
          7. Disc flowers 35–50 per head; ray flowers 15–20 per head . . . 8. *S. asteriscus*
        6. Cauline leaves in whorls of 3; basal leaves persistent at anthesis . . . .
        . . . . 7. *S. trifoliatum*

  1. **Silphium laciniatum** L. Sp. Pl. 2:919. 1753. Figs. 37a, 37b.

Stout perennial from a taproot; stems erect, terete, to 3.5 m tall, hirsute to hispid, very scabrous, resinous, mostly unbranched; leaves basal and cauline, the basal present at anthesis, pinnate or deeply pinnatifid, the divisions lanceolate to oblong, hispid to hirsute, harshly scabrous, on petioles up to 30 cm long; cauline leaves alternate, vertically aligned, pinnate or deeply pinnatifid, the divisions lanceolate to oblong, hispid to hirsute, harshly scabrous, to 60 cm long, to 30 cm wide, sessile or the lower ones short—petiolate or the upper ones cordate-clasping; heads several to numerous, to 10 cm across, on pubescent, glandular or eglandular peduncles; involucre hemispheric, 1.0–2.5 cm high; phyllaries in 2–3 series, acuminate to caudate at the apex, glabrous or glandular-pubescent or hispid, the outer lanceolate to ovate, stiff, usually somewhat spreading or reflexed, the inner usually appressed; receptacle flat, paleate; ray flowers up to 35, yellow, to 3.5 cm long, pistillate, fertile; disc flowers numerous, tubular, yellow, staminate, the corolla 5-lobed; cypselae oval, broadly winged, flat, deeply notched at the tip, 1.0–1.5 cm long, usually glabrous; pappus of awns 1–3 mm long.

  Two varieties occur in Illinois:

a. Peduncles and phyllaries eglandular . . . . 1a. *S. laciniatum* var. *laciniatum*
a. Peduncles and phyllaries glandular-pubescent . . . . 1b. *S. laciniatum* var. *robinsonii*

37a. *Silphium laciniatum*
var. *laciniatum*
(Compass plant).

a. Upper part of plant.
b, c. Phyllaries.
d. Flowering head.
e. Ray flower.
f. Palea.

g. Disc flower.
h. Cypsela.
i. Portion of plant with
   cauline leaves.
j. Leaf.

37a. (*continued*)
   Silphium
   laciniatum
   var. *laciniatum*
   (Compass plant).
i. Portion of plant
   with cauline leaves.
j. Leaf.

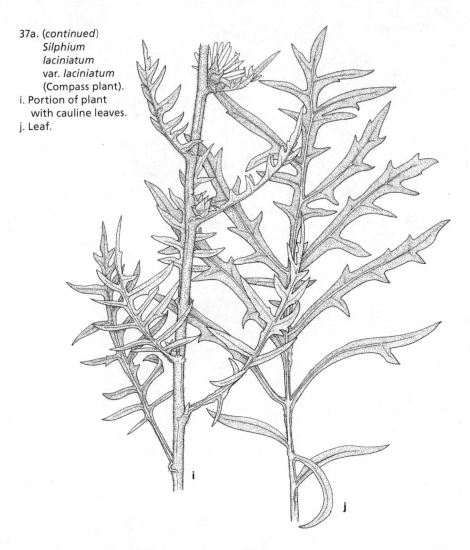

1a. **Silphium laciniatum** L. var. **laciniatum** Fig. 37a.
Peduncles and phyllaries eglandular.

Common Name: Compass plant.
Habitat: Prairies.
Range: New York to Ontario to Minnesota and South Dakota, south to New Mexico,
   Texas, and Alabama.
Illinois Distribution: Throughout the state, but less common in the southern
   counties.

The common name is derived from the fact that many of the vertical leaves are arranged so that their edges tend to be oriented north and south.

This plant differs from *S. pinnatifidum*, the other species of *Silphium* in Illinois with deeply pinnatifid leaves, by its acuminate to caudate phyllaries, the outermost often spreading or reflexed.

Compass plant flowers from June to September.

1b. **Silphium laciniatum** L. var. **robinsonii** L.M. Perry, Rhodora 39:247. 1937. Fig. 37b.

Silphium gummiferum Ell. Bot. S.C. & Ga. 2:460. 1823.

Peduncles and phyllaries glandular-pubescent.

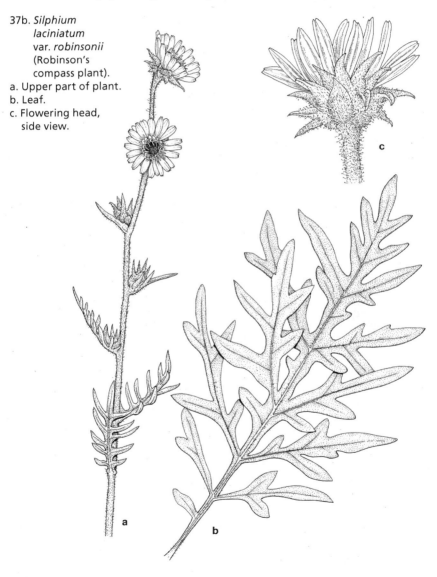

37b. *Silphium laciniatum* var. *robinsonii* (Robinson's compass plant).
a. Upper part of plant.
b. Leaf.
c. Flowering head, side view.

Common Name: Robinson's compass plant.
Habitat: Prairies.
Range: Indiana and Illinois, south to Oklahoma, Louisiana, and Alabama.
Illinois Distribution: Occasional in the northern three-fourths of Illinois.

This variety flowers from June to September.

2. **Silphium terebinthinaceum** Jacq. Hort. Bot. Vindob. 1:16. 1770. Figs. 38a, 38b.

Stout perennial herb from a taproot; stems erect, glabrous, branched above, to 3 m tall; leaves mostly basal, with a few alternate, cauline leaves, ovate to broadly oblong, acute at the apex, cordate or truncate at the base, usually harshly scabrous on both surfaces or sometimes more or less glabrous on the upper surface, to 40 cm long, about as wide, coarsely toothed or sometimes entire, petiolate; heads numerous, radiate, to 3.5 cm across, on glabrous peduncles; involucre hemispheric, to 4 cm across; phyllaries in (2–) 3 series, ovate-oblong, obtuse to subacute at the apex, appressed, glabrous or puberulent; receptacle flat, paleate; ray flowers up to 30, yellow, to 3.5 cm long, pistillate, fertile; disc flowers numerous, tubular, yellow, staminate, the corolla 5-lobed; cypselae obovate, flat, narrowly winged, slightly notched at the apex, glabrous or nearly so, to 12 mm long, to 10 mm wide; pappus absent or of awns less than 1 mm long.

Two varieties occur in Illinois:
a. Upper leaf surface harshly scabrous . . . . . 2a. *S. terebinthinaceum* var. *terebinthinaceum*
a. Upper leaf surface glabrous or nearly so. . . . . . 2b. *S. terebinthinaceum* var. *lucy-brauniae*

2a. **Silphium terebinthinaceum** Jacq. var. **terebinthinaceum** Fig. 38a.
Upper leaf surface harshly scabrous.

Common Name: Prairie dock.
Habitat: Prairies, glades, fens.
Range: New Hampshire and Ontario to Wisconsin and Nebraska, south to Arkansas, Missouri, and Georgia.
Illinois Distribution: Common throughout the state.

This is one of the more characteristic species of prairies in Illinois. It flowers from June to September.

2b. **Silphium terebinthinaceum** Jacq. var. **lucy-brauniae** Steyerm. Rhodora 53:134–135. 1951. Fig. 38b.
Upper leaf surface glabrous or nearly so.

Common Name: Lucy Braun's prairie dock.
Habitat: Prairies.
Range: Indiana, Illinois, Kentucky, and Missouri.
Illinois Distribution: Known from Cook and Jackson counties.

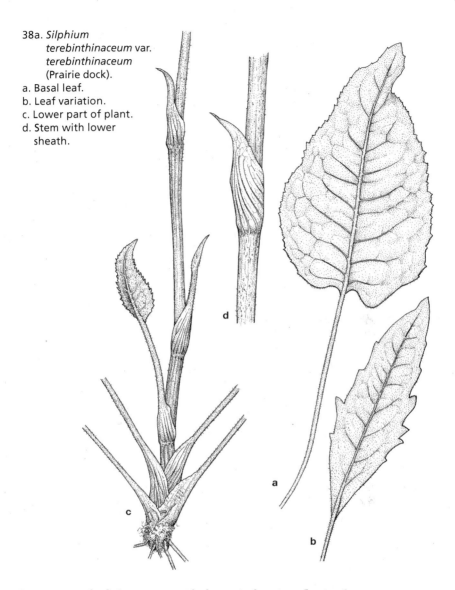

38a. *Silphium terebinthinaceum* var. *terebinthinaceum* (Prairie dock).
a. Basal leaf.
b. Leaf variation.
c. Lower part of plant.
d. Stem with lower sheath.

This variety, which may occur with the typical variety, flowers from June to September.

3. **Silphium pinnatifidum** Ell. Bot. S.C. & Ga. 2:462. 1823. Fig. 39.
Silphium terebinthinaceum Jacq. var. *pinnatifidum* (Ell.) Gray, Man. 220. 1848.

Perennial herb from a taproot; stems erect, glabrous, branched above, to 2.5 m tall; leaves mostly basal, with a few alternate cauline leaves, pinnately 5- to 13-lobed, the sinuses often deep, subacute at the apex, rounded at the base,

38a. (continued) *Silphium terebinthinaceum* var. *terebinthinaceum* (Prairie dock).

e. Upper part of plant.
f. Cluster of phyllaries.
g. Outer phyllary.
h. Inner phyllary.

i. Flowering head.
j. Ray flower.
k. Disc flower.

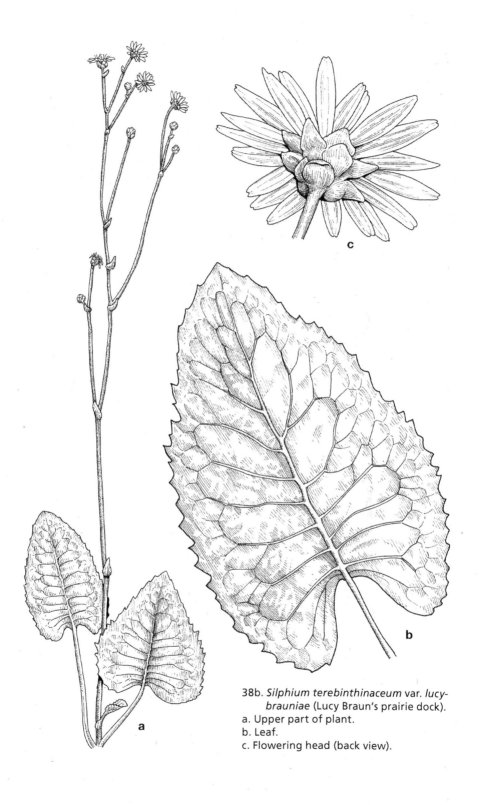

38b. *Silphium terebinthinaceum* var. *lucy-
   brauniae* (Lucy Braun's prairie dock).
a. Upper part of plant.
b. Leaf.
c. Flowering head (back view).

scabrous on both surfaces, to 50 cm long, nearly as wide, petiolate; heads several, radiate, to 3 cm across, on glabrous peduncles; involucre hemispheric, to 3.5 cm across; phyllaries in 2–3 series, appressed, oblong-ovate, obtuse to subacute at the apex, usually glabrous; receptacle flat, paleate; ray flowers up to 20, yellow, to 3 cm long, pistillate, fertile; disc flowers numerous, tubular, yellow, staminate, the corolla 5-lobed; cypselae flat, obovate, narrowly winged, usually glabrous, up to 12 mm long, up to 10 mm wide; pappus of a few awns less than 1 mm long.

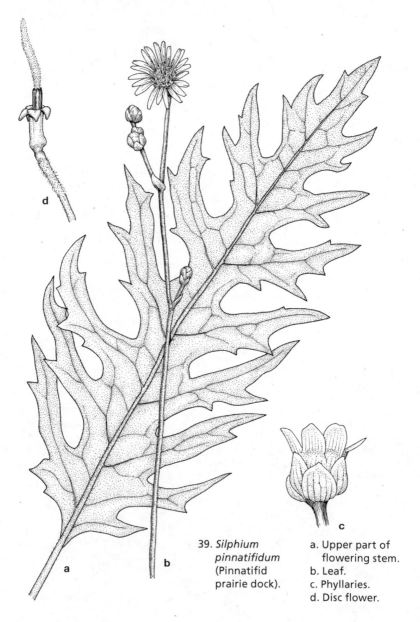

39. *Silphium pinnatifidum* (Pinnatifid prairie dock).

a. Upper part of flowering stem.
b. Leaf.
c. Phyllaries.
d. Disc flower.

Common Name: Pinnatifid prairie dock.
Habitat: Prairies.
Range: Ohio, Michigan, and Illinois, south to Alabama and Georgia.
Illinois Distribution: Scattered but rare in the state.

This very distinctive-appearing plant is usually considered to be a variety of *S. terebinthinaceum*. Its geographic range is much more restricted than that of *S. terebinthinaceum*.

The flowers of *S. pinnatifidum* bloom from June to September.

4. **Silphium perfoliatum** L. Syst. Nat., ed. 10, 2:1232. 1759. Fig. 40.
Robust perennial from fibrous roots; stems erect, often branched, 4-angled, glabrous or hispid, to 2.7 m tall; leaves opposite, rarely in whorls of 3, ovate to deltoid-ovate, acute at the apex, more or less truncate at the base, scabrous on both surfaces, coarsely dentate, to 40 cm long, to 25 cm wide, the upper connate-perfoliate at the base, the lower abruptly contracted to a broadly winged petiole; heads several to numerous, radiate, to 5 cm across, in corymbs, on glabrous or scabrous peduncles; involucre hemispheric or sometimes flattened; phyllaries in 2–3 series, the outer broadly ovate, acute at the apex, spreading to erect, ciliolate, scabrous to hispid; receptacle flat, paleate; ray flowers up to 30, to 2.2 cm long, yellow, pistillate, fertile; disc flowers numerous, tubular, yellow, staminate, the corolla 5-lobed; cypselae flat, obovate, winged, notched at the apex, glabrous, 10–12 mm long; pappus of teeth or awns up to 1.5 mm long.

Common Name: Cup-plant.
Habitat: Floodplain woods, along streams, wet ground.
Range: Vermont to Ontario to North Dakota, south to Texas, Alabama, and North
   Carolina.
Illinois Distribution: Common throughout the state.

This species is conspicuous by its square stem and its connate-perfoliate leaves, one of five species in Illinois to show this trait. The connate-perfoliate leaves form a cup where dew and rainwater collect.

Occasional specimens may be found that have some or all the leaves in whorls of three.

Silphium perfoliatum flowers from July to October.

5. **Silphium integrifolium** Michx. Fl. Bor. Am. 2:146. 1803. Figs. 41a, 41b, 41c.
Perennial herbs from fibrous roots; stems erect, branched above, terete or slightly angular, glabrous or occasionally scabrous, to 2 m tall; basal leaves withering early; cauline leaves opposite, ovate to lanceolate, acute to acuminate at the apex, tapering or rounded at the base, to 17 cm long, to 8 cm wide, usually harshly scabrous, hispid, rarely nearly glabrous, or the lower surface sometimes velvety-pubescent, entire to finely serrate, sessile; heads numerous in corymbs, radiate, usually on hispidulous peduncles; involucre hemispheric; phyllaries in 2–3 series,

40. *Silphium*
    perfoliatum
    (Cup-plant).
a. Upper part of plant.

b, c. Node with base of
    connate leaves.
d. Phyllaries.
e. Phyllary.
f. Flowering head.

g. Palea.
h. Ray flower.
i. Palea.
j. Disc flower with palea.
k. cypsela.

the outer spreading or reflexed, narrowly ovate to ovate, erect, hispid, glandular or eglandular; receptacle flat, paleate; ray flowers up to 20 (–22), to 2 cm long, yellow, pistillate, fertile; disc flowers up to 130 (–140), tubular, yellow, staminate, the corolla 5-lobed; cypselae flat, obovate, deeply notched at the apex, narrowly winged, glabrous, 8–12 mm long; pappus of teeth or awns 1–4 mm long.

Three varieties occur in Illinois:

a. Phyllaries eglandular. . . . . 5a. *S. integrifolium* var. *integrifolium*
a. Phyllaries stipitate-glandular.
    b. Leaves velvety-pubescent on the lower surface . . . . 5b. *S. integrifolium* var. *deamii*
    b. Leaves hispid to hirsute on the lower surface. . . . . . . 5c. *S. integrifolium* var. *neglectum*

5a. **Silphium integrifolium** Michx. var. **integrifolium** Fig. 41a.
Phyllaries eglandular.

Common Name: Rosinweed.
Habitat: Prairies, sand prairies, prairie fens.
Range: Ontario to Wyoming, south to New Mexico, Texas, and Alabama.
Illinois Distribution: Occasional to common throughout Illinois.

This typical variety differs from var. *deamii* and var. *neglectum* by having eglandular phyllaries. It flowers during July and August.

5b. **Silphium integrifolium** Michx. var. **deamii** L.M. Perry, Rhodora 39:287. 1937. Fig. 41b.
Phyllaries stipitate-glandular; lower surface of leaves velvety-pubescent.

Common Name: Deam's rosinweed.
Habitat: Prairies, black oak savannas.
Range: Indiana, Wisconsin, and Iowa, south to Arkansas, Mississippi, and Alabama.
Illinois Distribution: Common throughout the state.

This variety flowers during July and August.

5c. **Silphium integrifolium** Michx. var. **neglectum** Settle & Fisher, Rhodora 72:540. 1970. Fig. 41c.
Phyllaries stipitate-glandular; lower surface of leaves hispid to hirsute.

Common Name: Neglected rosinweed; bald rosinweed.
Habitat: Prairies, bur oak savannas.
Range: Michigan to South Dakota, south to Texas and Alabama.
Illinois Distribution: Common throughout the state.

This variety flowers in July and August.

6. **Silphium speciosum** Nutt. Trans. Am. Phil. Soc., n.s., 7:341. 1840. Fig. 42.
Silphium integrifolium Michx. var. *laeve* Torr. & Gray, Fl. N. Am. 2:279. 1842.

41a. *Silphium integrifolium* a. Upper part of plant.   c. Ray flower with cypsela.
     var. *integrifolium* b. Disc flower.   d. Phyllaries.
     (Rosinweed).

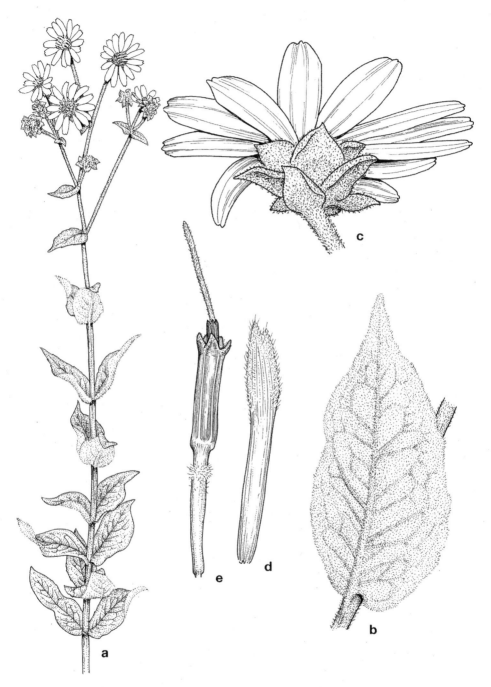

41b. *Silphium integrifolium* var. *deamii* (Deam's rosinweed).

a. Upper part of plant.
b. Leaf.
c. Flowering head.
d. Palea.
e. Disc flower.

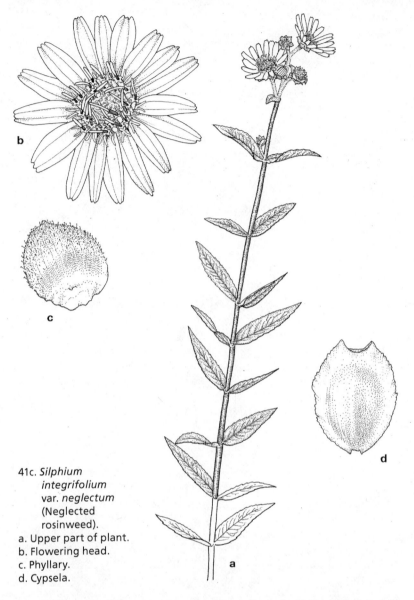

41c. *Silphium integrifolium* var. *neglectum* (Neglected rosinweed).
a. Upper part of plant.
b. Flowering head.
c. Phyllary.
d. Cypsela.

Perennial herb from fibrous roots; stems erect, branched above, slightly angular, glaucous, glabrous, to 1.5 m tall; basal leaves withering early; cauline leaves opposite, narrowly ovate to ovate, acute at the apex, tapering or rounded at the base, to 15 cm long, to 7.5 cm wide, glabrous, usually glaucous, entire to denticulate, sessile; heads several to numerous in corymbs, radiate, on glabrous peduncles; involucre hemispheric; phyllaries in 2–3 series, the outer usually reflexed, glabrous, the inner erect, glabrous; receptacle flat, paleate; ray flowers 20 or more, to 1.8 cm long, yellow, pistillate, fertile; disc flowers 130 or more, tubular, yellow, staminate, the corolla 5-lobed; cypselae flat, obovate, glabrous, notched at the apex, narrowly winged, 8–12 mm long; pappus of teeth or awns up to 3 mm long.

Common Name: Smooth rosinweed.
Habitat: Along a railroad (in Illinois).
Range: Native to the western United States; adventive in Illinois.
Illinois Distribution: Known only from Lake County where it was collected along a
  railroad.

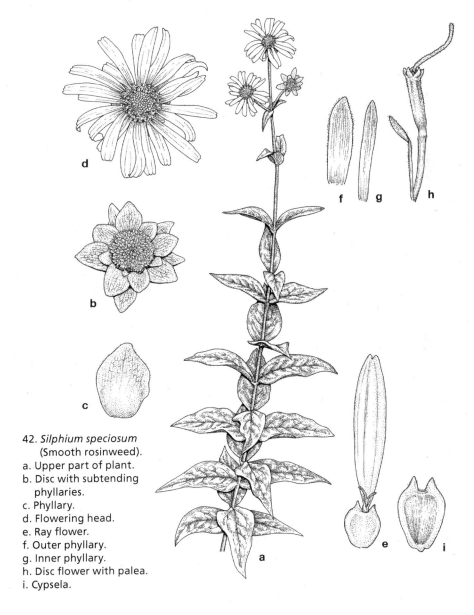

42. *Silphium speciosum*
    (Smooth rosinweed).
a. Upper part of plant.
b. Disc with subtending
   phyllaries.
c. Phyllary.
d. Flowering head.
e. Ray flower.
f. Outer phyllary.
g. Inner phyllary.
h. Disc flower with palea.
i. Cypsela.

This species, sometimes considered to be a variety of *S. integrifolium*, differs by its glaucous and glabrous stems and leaves, its glabrous peduncles and phyllaries, its more numerous ray flowers, and its more numerous disc flowers.

Silphium speciosum flowers from July to September.

7. Silphium trifoliatum L. Sp. Pl. 2:290. 1753. Fig. 43.
Silphium laevigatum Pursh, Fl. Am. Sept. 2:578. 1814.
Silphium asteriscus L. var. *trifoliatum* (L.) J.A. Clev. Novon 14:276. 2004.

Perennial herbs from fibrous roots; stems erect, branched above, terete, glabrous, sometimes glaucous, to 3 m tall; basal leaves persistent, lanceolate to lance-ovate, acuminate at the tip, tapering to the base, entire or finely serrate, to 20 cm long, to 5 cm wide, sparsely pubescent on both surfaces; cauline leaves in whorls of 3 or less commonly 4, or a few of them opposite or alternate, lanceolate, acuminate at the tip, tapering to the sessile or short-petiolate base, entire or finely serrate, to 15 cm long, to 4 cm wide, sparsely pubescent on both surfaces; heads several to numerous in corymbs, radiate, to 3.5 cm across, on glabrous peduncles; involucre hemispheric, glabrous; phyllaries in 2–3 series, the outer ovate to oval, acute or obtuse at the apex, glabrous, spreading, the inner broadly elliptic, erect; receptacle flat, paleate; ray flowers 15–20, yellow, to 1.8 cm long, pistillate, fertile; disc flowers 35–90, tubular, yellow, staminate, the corolla 5-lobed; cypselae flat, obovate, notched at the apex, narrowly winged, 6–10 mm long, glabrous or nearly so; pappus of a few teeth or awns up to 5 mm long.

Common Name: Whorled rosinweed.
Habitat: Dry woods.
Range: New York to Illinois, south to Alabama and Georgia.
Illinois Distribution: Known only from Hardin County.

This species is readily recognized by its middle and upper cauline leaves that are in whorls of three or less commonly four. It is sometimes considered to be a variety of *S. asteriscus*, but it seems to me that there are several significant differences. The following chart summarizes the differences:

	S. trifoliatum	*S. asteriscus*
basal leaves	persistent	caducous
stems	glabrous	hispid
cauline leaves	some in whorls of 3 or 4	not whorled
	some opposite or alternate	all opposite
	acuminate at apex	acute at apex
	sparsely pubescent	hispid
	short-petiolate	sessile
peduncle	glabrous	hispid
ray flowers	15–20	11–15

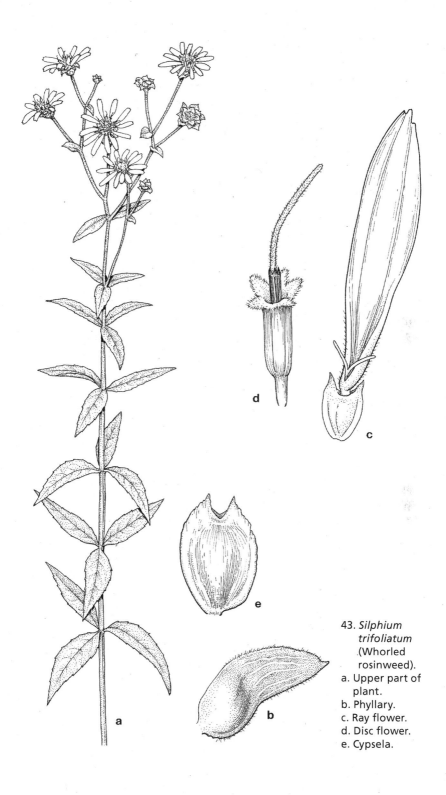

43. *Silphium trifoliatum* (Whorled rosinweed).
a. Upper part of plant.
b. Phyllary.
c. Ray flower.
d. Disc flower.
e. Cypsela.

The Illinois specimen was collected in a rocky woods in Hardin County.

Silphium trifoliatum flowers from July to September.

8. **Silphium asteriscus** L. Sp. Pl. 2:920. 1753. Fig. 44.

Perennial herbs from fibrous roots; stems erect, branched above, terete, hispid, to 3 m tall; basal leaves withered at flowering time, lanceolate to lance-ovate, acuminate at the apex, tapering to the base, entire or finely serrate, to 20 cm long, to 5 cm wide, hispid on both surfaces, the cauline leaves opposite, lanceolate, acute at the apex, tapering to the sessile base, entire or finely serrate, to 15 cm long, to 4 cm wide, hispid; heads several to numerous in corymbs, radiate, to 3.5 cm across, on hispid peduncles; involucre hemispheric; phyllaries in 2–3 series, the outer ovate to oval, acute or obtuse at the apex, glabrous, spreading, the inner broadly elliptic, erect; receptacle flat, paleate; ray flowers 11–15, yellow, to 1.8 cm long, pistillate, fertile; disc flowers 35–90, tubular, yellow, staminate, the corolla 5-lobed; cypselae flat, obovate, notched at the apex, narrowly winged, 6–10 mm long, glabrous or nearly so; pappus of a few teeth or awns up to 5 mm long.

Common Name: Starry rosinweed.
Habitat: Rocky glade.
Range: Virginia to Missouri, south to Texas and Florida.
Illinois Distribution: Known only from Union County.

This species resembles *S. integrifolium*, differing by its fewer ray and disc flowers per head.

See above for differences between *S. asteriscus* and *S. trifoliatum*.

Silphium asteriscus flowers from July to September.

44. **Echinacea** Moench—Purple Coneflower

Perennial herbs from taproots or fibrous roots; stems erect, branched or unbranched; leaves basal and cauline, the basal petiolate, 3- or 5-nerved, the cauline alternate, simple, undivided; head radiate, borne singly on a long peduncle; involucre hemispheric but usually somewhat flattened; phyllaries in 2–4 series, reflexed, spreading, or appressed, equal or nearly so; receptacle usually conic, paleate, the paleae longer than the corolla of the disc flowers; ray flowers up to 20, purple or pink (in our species), neutral, often persistent; disc flowers numerous, perfect, fertile, tubular, the corolla 5-lobed; cypselae usually 4-angled, thick; pappus a crown of up to 4 teeth.

Nine species comprise this genus, all of them in the eastern and central United States.

Four species occur in Illinois:
1. Leaves lanceolate to linear, tapering to the base, entire.
   2. Ray flowers 15–40 mm long.....................................1. *E. angustifolia*
   2. Ray flowers 40–90 mm long.
      3. Pollen white; rays 3–4 mm wide; cypselae glabrous.................2. *E. pallida*
      3. Pollen yellow; rays 4–7 mm wide; cypselae pubescent ....3. *E. simulata*

44. *Silphium asteriscus*
 (Starry rosinweed).
a. Upper part of plant.
b. Leaf.

c. Cluster of phyllaries.
d. Phyllary.
e. Flowering head.
f. Ray flower.

g. Palea.
h. Disc flower.
i. Cypsela.

1. Leaves ovate, rounded at the base, serrate or dentate, or the upper frequently entire...
....4. *E. purpurea*

1. Echinacea angustifolia DC. Prodr. 5:554. 1836. Fig. 45.
Brauneria angustifolia (DC.) Heller, Muhlenbergia 1:5. 1900.

Perennial herbs from branched taproots; stems erect, unbranched, hispid to hirsute, the hairs tuberculate-based, to 75 cm tall; leaves linear-lanceolate to

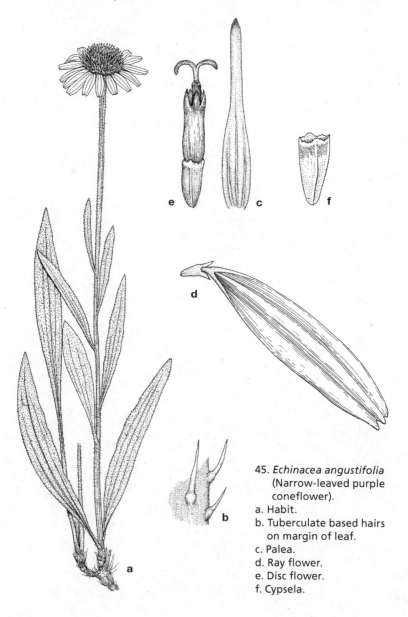

45. *Echinacea angustifolia*
    (Narrow-leaved purple
    coneflower).
a. Habit.
b. Tuberculate based hairs
   on margin of leaf.
c. Palea.
d. Ray flower.
e. Disc flower.
f. Cypsela.

lanceolate, acute at the apex, tapering to the base, to 30 cm long, to 3.5 cm wide, hirsute, entire, ciliolate along the margins, strongly 3-nerved, with another pair of less distinct nerves, the basal and lower cauline leaves on petioles up to 2 cm long, the upper cauline leaves smaller and on shorter petioles, or sessile; head radiate, borne singly on a hispid peduncle up to 30 cm long; involucre depressed-hemispheric; phyllaries in 2–4 series, equal, lanceolate to ovate, to 12 mm long; receptacle conic, paleate, the paleae to 15 mm long; ray flowers up to 16 (–18), purple, notched at the tip, spreading, 15–40 mm long, neutral; disc flowers numerous, perfect, fertile, tubular, the corolla 5-lobed, 5–7 mm long; cypselae 4-angled, obpyramidal, glabrous or nearly so, tan and brown, 4–5 mm long; pappus a crown of teeth up to 1 mm long.

Common Name: Narrow-leaved purple coneflower.
Habitat: Hill prairies.
Range: Manitoba and Saskatchewan, south to New Mexico, Texas, and Illinois.
Illinois Distribution: Scattered in a few western counties, but very rare.

This species, which apparently is confined in Illinois to hill prairies along the Mississippi River, has purple ray flowers similar in color to those of *E. purpurea*, but narrow leaves similar to those of *E. pallida*. The rays are spreading in *E. angustifolia* and usually drooping in *E. purpurea*. The hairs on the stems and leaves of *E. angustifolia* are tuberculate-based, while the hairs of *E. pallida* lack the tuberculate base.

  Echinacea angustifolia flowers from June to September.

  2. **Echinacea pallida** (Nutt.) Nutt. Trans. Am. Phil. Soc., n.s. 7:354. 1840. Fig. 46.
Rudbeckia pallida Nutt. Journ. Acad. Nat. Sci. Phil. 7:77. 1834.
Brauneria pallida (Nutt.) Britt. Mem. Torrey Club 5:333. 1894.

  Perennial herbs from branched taproots; stems erect, usually unbranched, hirsute, with non-tuberulate-based hairs, to 1.2 m tall; leaves linear-lanceolate to lanceolate, acuminate at the apex, tapering to the base, hirsute, entire, ciliolate on the margins, to 40 cm long, to 4 m wide, strongly 3- or 5-nerved, the basal on petioles to 20 cm long, the cauline on shorter petioles or sessile; head radiate, borne singly on a hispid peduncle up to 50 cm long; involucre depressed-hemispheric; phyllaries in 2–4 series, equal, lanceolate to ovate, to 15 mm long; receptacle conic, paleate, the paleae to 15 mm long; rays up to 16 (–18), pink, strongly reflexed, notched at the tip, 40–90 mm long, 3–4 mm wide, neutral; disc flowers numerous, pink or purplish, tubular, perfect, fertile, the corolla 5-lobed, 5.5–7.0 mm long; pollen usually white; cypselae 4-angled, obpyramidal, glabrous, tan or tan and brown, 2.5–5.0 mm long; pappus a crown of teeth about 1 mm long.

Common Name: Pale purple coneflower.
Habitat: Dry prairies, open woods, limestone glades.
Range: Maine to Ontario, south to Nebraska, Texas, Alabama, and North Carolina.
Illinois Distribution: Occasional throughout Illinois.

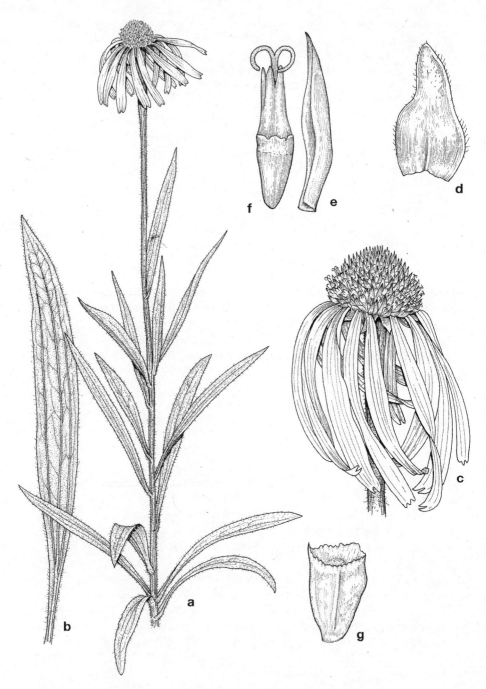

46. *Echinacea pallida* (Pale purple coneflower).
a. Upper part of plant.
b. Leaf.
c. Flowering head.
d. Phyllary.
e. Palea.
f. Disc flower.
g. Cypsela.

The rays are usually very pale pink and strongly reflexed. The narrow leaves are similar in shape to those of *E. angustifolia*, but the hairs are not tuberculate-based. Its rays are much longer than those of *E. angustifolia*. *Echinacea pallida* is also very similar to *E. simulata*, but the pollen is usually white, the rays are narrower, and the cypselae are glabrous, whereas *E. simulata* has yellow pollen, wider rays, and pubescent cypselae.

The flowers bloom from April to May.

3. **Echinacea simulata** McGregor, Sida 3:282. 1968. Fig. 47.
Echinacea speciosa McGregor, Trans. Kans. Acad. Sci. 70:366. 1967, *non* (Wenderoth) Paxton (1849).
Echinacea pallida (Nutt.) Nutt. var. *simulata* (McGregor) Birns, B.R. Baum. & Amason, Syst. Bot. 27:629. 2002.

Perennial herbs from branched taproots; stems erect, usually unbranched, sparsely to densely hirsute, to 1 m tall; basal and cauline leaves linear to lanceolate, acute to acuminate at the apex, tapering to the base, entire, hirsute, ciliolate along the margins, 3- or 5-nerved, to 40 cm long, to 4 cm wide, the basal long-petiolate, the middle and upper cauline short-petiolate or sessile; head radiate, borne singly on a hispid peduncle up to 40 cm long; involucre hemispheric; phyllaries in 2–4 series, equal, lanceolate to ovate, to 15 mm long; receptacle conic, paleate, the paleae up to 12 (–14) mm long; rays up to 20, pink, recurved to reflexed, up to 9 cm long, 4–7 mm wide, neutral; disc flowers numerous, pink or purplish, tubular, perfect, fertile, the corolla 5-lobed, 5.0–6.5 mm long; pollen yellow; cypselae 4-angled, obpyramidal, sparsely pubescent, tan, 3.0–4.5 mm long; pappus a crown of teeth up to 1 mm long.

Common Name: Wavy-leaf purple coneflower.
Habitat: Open woods, dry prairies.
Range: Kentucky to Illinois and Missouri, south to Arkansas, Tennessee, and Georgia.
Illinois Distribution: Scattered in Illinois, but not common.

This species is closely related to *E. pallida*, but differs by its wider rays, sparsely pubescent cypselae, and yellow pollen.

Echinacea simulata flowers during July and August.

4. **Echinacea purpurea** (L.) Moench, Meth. 591. 1794. Fig. 48.
Rudbeckia purpurea L. Sp. Pl. 2:907. 1753.
Brauneria purpurea (L.) Britt. Mem. Torrey Club 5:334. 1894.

Perennial herbs from fibrous roots; stems erect, rather stout, usually unbranched, glabrous or less commonly sparsely hispid, to 1.5 m tall; leaves ovate to ovate-lanceolate, acute to acuminate at the apex, usually rounded at the base, serrate or dentate, to 30 cm long, to 12 cm wide, strongly 3- or 5-nerved, glabrous

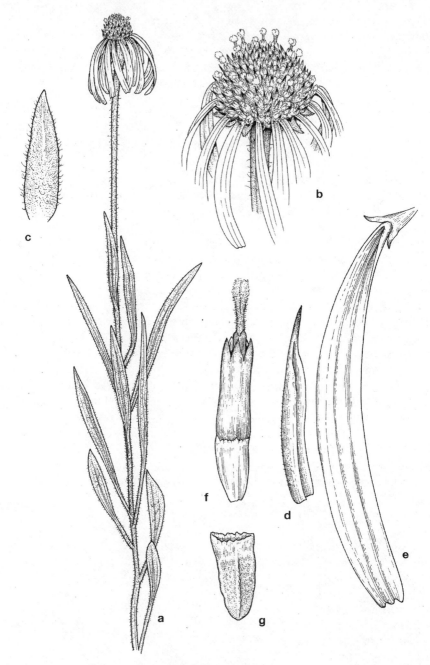

47. *Echinacea simulata* (Wavy-leaf purple coneflower).

a. Upper part of plant.
b. Flowering head.
c. Phyllary.
d. Palea.

e. Ray flower.
f. Disc flower.
g. Cypsela.

or sparsely hispid, the basal on petioles up to 20 cm long, the middle and upper cauline usually narrower, sometimes entire, short-petiolate to nearly sessile; head radiate, borne singly on a usually glabrous peduncle to 25 cm long; involucre hemispheric; phyllaries in 2–4 series, equal, linear to lanceolate, to 15 (–17) mm long; rays up to 20, purple, spreading to recurved, up to 7.5 cm long, neutral; disc flowers numerous, pinkish or purple, tubular, perfect, fertile, the corolla 5-lobed, 4.5–6.0 mm long; cypselae 4-angled, obpyramidal, pale, glabrous or sparsely pubescent, 3.5–5.0 mm long; pappus a crown of teeth 1.0–1.2 mm long.

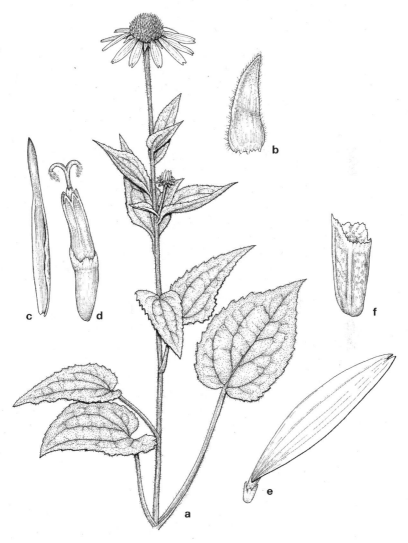

48. *Echinacea purpurea* (Purple coneflower).

a. Upper part of plant.
b. Phyllary.
c. Palea.
d. Disc flower.
e. Ray flower.
f. Cypsela.

Common Name: Purple coneflower; broad-leaved coneflower.
Habitat: Prairies, open woods, wooded floodplains.
Range: Ontario to Wisconsin and Kansas, south to Texas and Illinois.
Illinois Distribution: Occasional throughout Illinois.

This is the only species of *Echinacea* in Illinois with ovate leaves. Its purple rays are either spreading or somewhat reflexed.

Echinacea purpurea is a popular ornamental and a necessity in any prairie wildflower seed mix.

This species flowers from July to September.

45. **Verbesina** L., *nomen conserv.*—Crownbeard; Wingstem

Perennial herbs from rhizomes; stems erect, narrowly winged; leaves alternate or occasionally opposite, simple; heads up to 100 in corymbs, radiate; involucre campanulate, 3–5 mm across; phyllaries 8–12 in 1 or 2 series, erect or spreading; receptacle flat to conic, paleate; ray flowers 1–5, white or yellow, pistillate; disc flowers 8–13, white or yellow, tubular, perfect, fertile, the corolla 5-lobed; cypselae flat, winged or wingless; pappus of 2 awns, or absent.

Over a hundred species, mostly in the New World, comprise this genus.

In the past, there have been five species of *Verbesina* attributed to Illinois. After years of studying the *Verbesina* complex, I have concluded that two of these five should be considered in the genus *Verbesina*, two considered to be in the genus *Actinomeris*, and one considered to be in the genus *Ximenesia*. As treated here, *Verbesina* consists of perennials with up to 100 flowering heads per plant, narrowly winged stems, phyllaries 8–12 per head and erect or spreading, campanulate involucres 3–5 mm across, white or yellow pistillate ray flowers with 1–5 rays, 8–13 white or yellow disc flowers, winged or wingless cypselae, and pappus of 2 awns or absent. *Actinomeris*, as treated here, consists of perennials with 1–50 flowering heads per plant, winged stems, phyllaries 8–12 per head and erect or deflexed, hemispheric involucres 10–15 mm across, yellow and usually neutral ray flowers with (2–) 5–13 rays, 40–100 yellow disc flowers, broadly winged cypselae, and pappus of 2–3 awns. *Ximenesia* consists of annuals with 1–3 flowering heads per plant, wingless stems, phyllaries 12–18 per head and spreading, hemispheric involucres 10–20 mm across, yellow pistillate ray flowers with 12–15 rays notched at the tip, 80–150 yellow disc flowers, both winged and unwinged cypselae, and pappus of 2 awns or sometimes absent.

Linnaeus apparently had different concepts of *Verbesina*. Although Linnaeus described *Verbesina*, he included not only *V. virginica* but also the plant we know as *Eclipta prostrata*. Our *Verbesina occidentalis* was described by Linnaeus as *Siegesbeckia occidentalis*, and the plant known by many as *Verbesina alternifolia* or *Actinomeris alternifolia* Linnaeus called *Coreopsis alternifolia*.

Two species of *Verbesina* occur in Illinois:

1. Most leaves alternate; rays white, 3–7 mm long; phyllaries in 1 series; disc flowers white; cypselae flat from the ray flowers, 3-angled from the disc flowers . 1. *V. virginica*
1. Most leaves opposite; rays yellow, 10–20 mm long; phyllaries in 2 series; disc flowers yellow; all cypselae flat . . . . 2. *V. occidentalis*

1. **Verbesina virginica** L. Sp. Pl. 2:901. 1753. Fig. 49.
Phaethusa virginica (L.) Britt. Ill. Fl. N.U.S. ed. 2, 3:487. 1913.

Perennial herbs from rhizomes; stems erect, branched or unbranched, densely downy pubescent, narrowly winged, to 2.5 m tall; leaves alternate, or the lowermost sometimes opposite, lance-ovate, acute at the apex, tapering to the base, to 10 cm long, to 6 cm wide, serrate, pubescent, scabrous on the upper surface, on

49. *Verbesina virginica* (Frostweed).
a. Upper part of plant.
b. Winged stem.
c. Cluster of flowering heads.
d. Phyllary.
e. Flowering head.
f. Ray flower.
g. Disc flower with subtending palea.
h. Cypsela.

short, winged petioles; heads radiate, up to 100, in corymbs; involucre campanulate, 3–5 mm across; phyllaries 8–12, in 1 series, appressed, to 5 (–7) mm long; receptacle low-conic, paleate; ray flowers 3–5, white, 3–7 mm long, pistillate; disc flowers 8–12 (–15), white, perfect, fertile, the corolla 5-lobed; cypselae of the ray flowers 3-sided, of the disc flowers flat, oblanceolate, winged, 3.5–5.0 mm long, puberulent; pappus of two awns up to 3 mm long, or occasionally absent.

Common Name: Frostweed; white crownbeard; tickseed.
Habitat: Dry open woods.
Range: Virginia to Ohio to Kansas, south to Texas and Florida.
Illinois Distribution: Not common in the extreme southeastern counties of Illinois.

This is the only white-flowered species in the *Verbesina* complex in Illinois. The ray flowers, disc flowers, involucre, and pappus are very similar to those of *V. occidentalis*, except that *V. virginica* has white rays, a white disc, and alternate cauline leaves.

This species has been placed in the genus *Phaethusa* in the past, but *Verbesina* has been conserved by the International Association of Plant Taxonomists.

Verbesina occidentalis flowers during July and August.

2. **Verbesina occidentalis** (L.) Walt. Fl. Carol. 213. 1788. Fig. 50.
Siegesbeckia occidentalis L. Sp. Pl. 2:900. 1753.
Phaethusa americana Gaertn. Fr. & Sem. 2:425. 1791.
Verbesina siegesbeckia Michx. Fl. Bor. Am. 2:134. 1803.
Phaethusa occidentalis (L.) Britt. In Britt. & Brown, Ill. Fl., ed. 2, 3:488. 1913.

Perennial herbs from rhizomes; stems erect, branched above, narrowly 4-winged, to 2 m tall, glabrous or sometimes pubescent in the upper half; leaves opposite, simple, lance-ovate to ovate, acute to acuminate at the apex, tapering to the base, to 12 cm long, to 6 cm wide, serrate, 3-nerved, pubescent on both surfaces, scabrous above, on short petioles; heads radiate, up to 100, in corymbs; involucre campanulate, 3–5 mm across; phyllaries 8–12, in 2 series, appressed, to 7 mm long; receptacle flat, paleate; ray flowers 1–5, yellow, 10–20 mm long, pistillate; disc flowers 8–13, yellow, tubular, perfect, fertile, the corolla 5-lobed; cypselae flat, oblanceolate, 4–5 mm long, wingless, strigose; pappus of 2 awns.

Common Name: Small yellow crownbeard; opposite-leaved wingstem.
Habitat: Bottomland forest.
Range: Virginia to Ohio and Missouri, south to Texas and Florida.
Illinois Distribution: Very rare in the southernmost counties.

This species is recognized by its narrowly winged stem, phyllaries in 2 series, few yellow rays, and opposite leaves.

Linnaeus thought this species was different enough from his concept of *Verbesina* to describe it in the genus *Siegesbeckia*.

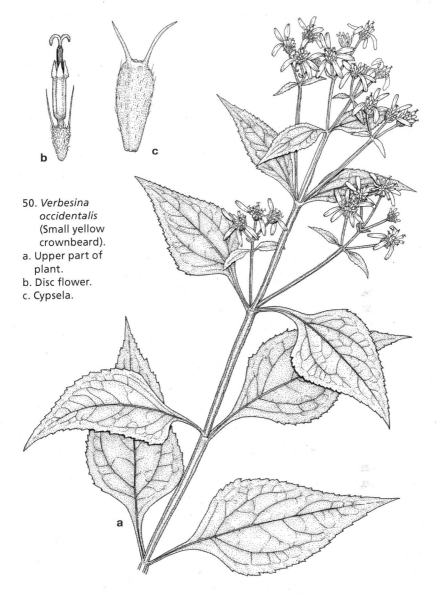

50. *Verbesina occidentalis* (Small yellow crownbeard).
a. Upper part of plant.
b. Disc flower.
c. Cypsela.

This species is an inhabitant of bottomland forests in the southernmost counties of the state.

Verbesina occidentalis flowers from August to October.

46. **Actinomeris** Nutt.—Wingstem

Perennial herbs from rhizomes; stems erect, usually winged; leaves mostly alternate, simple; heads radiate, up to 50 per plant, borne singly or in small corymbs; involucre hemispheric, usually 10 mm across; phyllaries 8–12, in 1–3 series, deflexed or erect; receptacle conic to globose, paleate; ray flowers up to 15, usually at least 15 mm long, yellow, neutral or rarely pistillate, usually shallowly 2-notched at the tip; disc flowers 40–100, yellow, tubular, perfect, fertile, the corolla 5-lobed; cypselae flat, broadly winged; pappus of 2–3 awns up to 2 mm long.

Actinomeris is often merged with *Verbesina*, but differs by its fewer flowering heads per plant, its smaller campanulate involucre, its more numerous ray flowers that are usually neutral, and its much fewer disc flowers.

After studying the *Verbesina* complex in the field for many years, I recognize two species in *Actinomeris* in Illinois.

1. Stems 2-winged; phyllaries reflexed; flowering heads up to 50; ray flowers 2–10; cypselae spreading in all directions, forming a globose fruiting structure . . . . 1. *A. alternifolia*
1. Stems 4-winged; phyllaries erect; flowering heads up to 10; ray flowers 8–13; cypselae not radiating in all directions, not forming a globose fruiting structure . . . . 2. *A. helianthoides*

1. **Actinomeris alternifolia** (L.) DC. Prodr. 5:575. 1836. Fig. 51.
Coreopsis alternifolia L. Sp. Pl. 2:909. 1753.
Actinomeris squarrosa Nutt. Gen. 2:181. 1818.
Verbesina alternifolia (L.) Britt. ex Kearney, Bull. Torrey Club 20:485. 1893.
Ridan alternifolius (L.) Britt. Ill. Fl. N.U.S. ed. 2., 3:487. 1913.

Perennial herbs from rhizomes; stems erect, branched above, usually narrowly winged, usually pubescent, to 2 m tall; leaves alternate, or the lowermost sometimes opposite, lance-ovate, acute to acuminate at the apex, tapering to the usually decurrent base, to 20 cm long, to 6 cm wide, serrate or rarely subentire, scabrous on both surfaces; heads radiate, up to 50 per plant, in corymbs; involucre hemispheric, 10–12 mm across; phyllaries 8–12, in 1–3 series, reflexed; receptacle globose, paleate; ray flowers 2–10, yellow, 15–25 mm long, neutral; disc flowers 40–100, yellow, tubular, perfect, the corolla 5-lobed; cypselae flat, oblanceolate, broadly winged, 4.5–5.0 mm long, sparsely pubescent; pappus of 2 (–3) subulate awns 0.5–2.0 mm long.

Common Name: Yellow ironweed; wingstem.
Habitat: Moist soil in open woods.
Range: New York and Ontario to Wisconsin and Nebraska, south to Texas and Florida.
Illinois Distribution: Occasional throughout Illinois.

This species is often placed in *Verbesina*, but it differs from that genus by its fewer flowering heads, its broader hemispheric involucre, its more numerous disc flowers, its broadly winged cypselae, and its globose fruiting structures.

This is a species of moist woods, flowering from August to October.

2. **Actinomeris helianthoides** (Michx.) Nutt. Gen. 2:181. 1818. Fig. 52.
Verbesina helianthoides Michx. Fl. Bor. Am. 2:135. 1803.

51. *Actinomeris alternifolia* (Yellow ironweed).

a. Upper part of plant.
b. Winged stem.
c. Phyllary.
d. Flowering head.
e. Ray flower.
f. Palea.
g. Disc flower.
h. Cypsela.

Phaethusa helianthoides (Michx.) Britt. Ill. Fl. N.U.S. ed. 2, 3:487. 1913.
Pterophyton helianthoides (Michx.) Alexander, Man. S.E. Fl. 1444. 1933.

Perennial herbs from rhizomes; stems erect, usually branched above, 4-winged, hispid to hirsute, to 1 m tall; leaves alternate, simple, lance-ovate to ovate, acute to acuminate at the tip, tapering to the sessile and usually decurrent base, to 10 cm long, to 4 cm wide, serrate, hispid and scabrous on the upper surface, soft-pubescent on the lower surface; heads radiate, 1–10 per plant, borne singly or in small corymbs; involucre hemispheric, 10–15 mm across; phyllaries 8–12, in 1–3 series, lanceolate, erect, to 9 mm long; receptacle conic, paleate; ray flowers 8–13, yellow, 20–30 mm long, pistillate or occasionally neutral; disc flowers 40–80, yellow, tubular, perfect, fertile, the corolla 5-lobed; cypselae flat, broadly winged, oblanceolate to elliptic, 4.5–5.0 mm long, pubescent; pappus of 2 subulate awns 0.5–1.5 mm long.

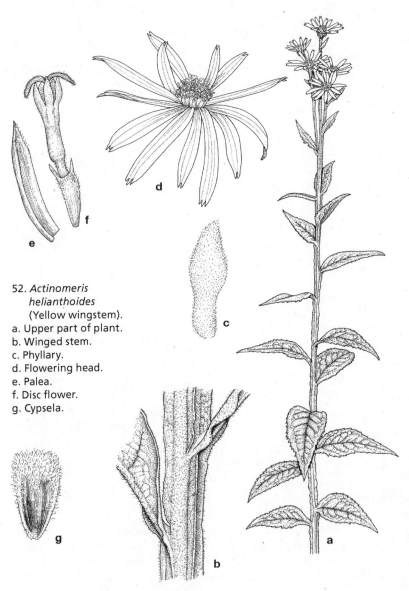

52. *Actinomeris*
    helianthoides
    (Yellow wingstem).
a. Upper part of plant.
b. Winged stem.
c. Phyllary.
d. Flowering head.
e. Palea.
f. Disc flower.
g. Cypsela.

Common Name: Yellow crownbeard; yellow wingstem; sunflower crownbeard.
Habitat: Open woods, prairies.
Range: Ohio to Iowa, south to Texas, Georgia, and North Carolina.
Illinois Distribution: Common in the southern three-fifths of the state, apparently
absent elsewhere.

This species is usually considered to be a species of *Verbesina*, but it appears to me
to have more common characteristics found in *Actinomeris*. It has fewer flowering
heads, hemispheric involucres at least 10 mm across, broadly winged cypselae,
and more ray flowers and more disc flowers than in *Verbesina*.

It differs from *A. alternifolia* by having erect rather than reflexed phyllaries, fewer
flowering heads, and usually more ray flowers per head. These differences have led
some botanists in the past to call it *Pterophyton helianthoides* (Michx.) Alexander.

Actinomeris helianthoides flowers earlier than other members of the *Verbesina*
complex, flowering from May to August.

47. **Ximenesia** Cav.—Golden Crownbeard

Annual herbs from fibrous roots; stems erect, much branched; leaves alternate, or
the lowest sometimes opposite, with or without dilated appendages at the base of
the petioles; flowering heads radiate, 1–3, showy, pedunculate; involucre hemispheric; phyllaries 12–18, in 1–2 series, spreading; receptacle convex, paleate; ray
flowers numerous, golden yellow, pistillate, fertile; disc flowers numerous, yellow,
tubular, perfect, fertile, the corolla 5-lobed; cypselae flat, winged; pappus of 2
subulate awns from the disc flowers and absent from the ray flowers.

For many years, this genus has been merged with *Verbesina*, but it differs by its
annual habit, unwinged stems, often a solitary flowering head, and 12 or more
spreading phyllaries per head.

1. **Ximenesia encelioides** Cav. Icon. 2:60. 1793. var. **exauriculata** (B.L. Rob. &
Greenm.) Mohlenb., **comb. nov.** (Basionym: *Verbesina encelioides* (Cav.) Benth. &
Hook. f. *exauriculata* B.L. Rob. & Greenm. Proc. Amer. Acad. Arts 34 (20):544.
1899). Fig. 53.
Verbesina encelioides (Cav.) Benth. & Hook. f. ex Gray in Brewer, Bot. Cal. 1:350.
1876.
Verbesina encelioides (Cav.) Benth. & Hook. f. *exauriculata* B.L. Rob. & Greenm. Proc.
Amer. Acad. Arts 34 (20):544. 1899.
Verbesina exauriculata (B.L. Rob. & Greenm.) Cockerell, Nature (London) 66:607.
1902.

Annual herbs from fibrous roots; stems erect, much branched, unwinged, canescent, to 1 m tall; leaves alternate, or the lowermost opposite, deltoid-lanceolate
to deltoid-ovate, acute at the apex, tapering or cordate at the base, to 10 cm long,
to 4 cm wide, coarsely dentate to nearly laciniate, pubescent on the upper surface,
canescent on the lower surface, without dilated appendages at the base of the petioles; heads 1–3, radiate, borne singly or in small corymbs; involucre hemispheric,

53. *Ximenesia encelioides*
   var. *exauriculata*
   (Golden crownbeard).

a. Upper part of plant.
b. Ray flower.

c. Disc flower.
d. Cypsela.

10–20 mm across; phyllaries 12–18, in 1–2 series, spreading, linear to lance-ovate, canescent, to 8 mm long; receptacle convex, paleate; ray flowers 12–15, golden yellow, 8–10 mm long, 3-notched at the apex, pistillate, fertile; disc flowers 80–150, yellow, tubular, perfect, fertile, the corolla 5-lobed; cypselae flat, obovate, broadly winged on the disc flowers, usually narrowly winged on the ray flowers, 3.5–5.0 mm long, rugose, slightly scabrous; pappus of disc flowers of 2 subulate awns 0.5–2.0 mm long, of ray flowers usually absent.

Common Name: Golden crownbeard.
Habitat: Disturbed soil (in Illinois).
Range: Native to the southwestern United States and Mexico; mostly introduced in
the eastern half of the United States.
Illinois Distribution: Known from Champaign, Madison, and St. Clair counties.

Our plants, which lack dilated appendages at the base of the petioles, may be
known as var. *exauriculata*. Typical var. *encelioides* has dilated appendages at the
base of the petioles.

Most botanists include this species within *Verbesina*, but it differs from other
species of *Verbesina* by its annual habit, unwinged stems, 1–3 flowering heads, and
12–18 spreading phyllaries.

Ximenesia helianthoides, which is sometimes planted as a garden ornamental, flowers from June to August. Its large flowering heads resemble those of a
sunflower.

48. **Melanthera** Rohr—Snow Squarestem

Perennials or subshrubs; stems ascending to erect, sometimes sprawling, usually
4-angled; leaves opposite, simple, 3-nerved, sometimes 3-lobed; heads discoid (in
our species), borne singly or several in corymbs; involucre hemispheric; phyllaries
in 2–3 series, unequal, persistent; receptacle flat to convex, paleate; ray flowers
absent (in our species); disc flowers numerous, white, tubular, perfect, fertile, the
corolla 5-lobed; cypselae obpyramidal, usually 4-angled; pappus of 2–12 caducous
barbellate bristles or awns.

Twenty species in the New World comprise this genus. Only the following occurs in Illinois.

1. **Melanthera nivea** (L.) Small, Fl. S.E.U.S. 1251. 1903. Fig. 54.
Bidens nivea L. Sp. Pl. 2:833. 1753.
Melanthera hastata Michx. Fl. Bor. Am. 2:107. 1803.

Perennial herb from rhizomes; stems ascending to erect, branched above, hispid,
4-sided, to 2 m tall; leaves opposite, simple, ovate to deltoid, acute to acuminate at
the apex, truncate or tapering to the petiolate base, to 12 cm long, to 8 cm wide,
serrate, hispid or strigose; heads discoid, several in terminal and axillary corymbs;
involucre hemispheric, 12–20 mm across; phyllaries 12–18, in 2–3 series, lanceolate to ovate, to 10 mm long; receptacle flat to convex, paleate, the paleae to 7 mm
long; ray flowers absent; disc flowers numerous, white, tubular, perfect, fertile, the
corolla 5-lobed, to 10 mm long; cypselae obpyramidal, 4-angled, brown, usually
glabrous, 2–3 mm long; pappus of 2–12 caducous, barbellate bristles.

Common Name: Snow squarestem.
Habitat: Floodplain woods.
Range: South Carolina to Louisiana and Florida; Illinois; Kentucky.
Illinois Distribution: Known only from Massac and Pulaski counties.

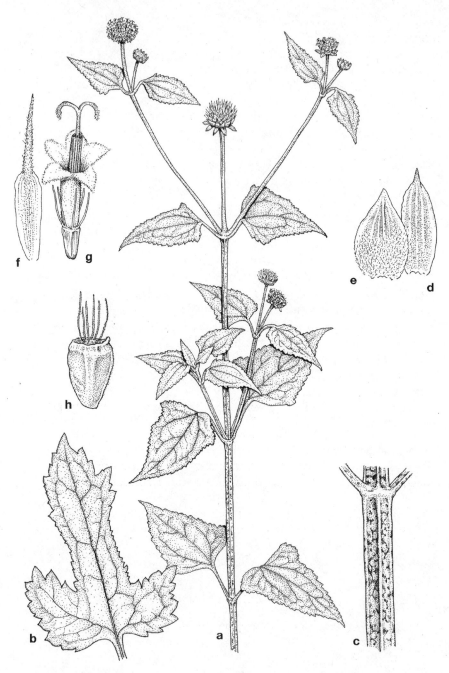

54. *Melanthera nivea*
(Snow squarestem).
a. Upper part of plant.
b. Leaf.

c. Section of square
stem.
d. Outer phyllary.
e. Inner phyllary.

f. Palea.
g. Disc flower.
h. Cypsela.

This attractive plant is distinguished by its snow-white discoid flowering heads, its opposite leaves, and its square stem. The sometimes similar-appearing *Ageratina altissima* lacks the square stem.

The Illinois locations are at the northernmost edge of the range of this species. *Melanthera nivea* flowers from June to October.

49. **Eclipta** L. *nomen conserv.*—Yerba de Tajo

Annual or perennial herbs; stems erect or decumbent, branched; leaves opposite, simple; heads radiate, borne singly or several in small terminal or axillary corymbs; involucre hemispheric; phyllaries in 2–3 series, equal, foliaceous, persistent; receptacle flat, paleate; ray flowers up to 40, pistillate, fertile; disc flowers up to 30, tubular, perfect, fertile, the corolla 4-lobed; cypselae of ray flowers 3- or 4-angled, of disc flowers flat; pappus a crown of persistent short teeth, or absent.

Four species comprise this genus, all in the New World.

Eclipta has been conserved since Linnaeus originally described our species in the genus *Verbesina*.

1. **Eclipta prostrata** (L.) L. Mant. Pl. 286. 1771. Fig. 55.
Verbesina prostrata L. Sp. Pl. 2:902. 1753.
Verbesina alba L. Sp. Pl. 2:902. 1753.
Eclipta erecta L. Mant. Pl. 286. 1771.
Eclipta procumbens Michx. Fl. Bor. Am. 2:129. 1803.
Eclipta brachypoda Michx. Fl. Bor. Am. 2:130. 1803.
Eclipta alba (L.) Hassk. Pl. Jav. Rar. 528. 1848.
Eclipta procumbens Michx. var. *brachypoda* (Michx.) Gray, Man. Bot. 218. 1848.

Annual herb from fibrous roots; stems procumbent to ascending, mauve-colored, with appressed white strigose hairs, to 1.5 m tall but usually much shorter; leaves opposite, simple, linear-lanceolate to lanceolate, acute to acuminate at the apex, tapering to the sessile base, to 10 cm long, to 3 cm wide, usually obscurely serrate, pubescent; head radiate, borne singly, on an appressed-pubescent peduncle; involucre hemispheric; phyllaries 8–12, in 2 rows, ovate-lanceolate, foliaceous, persistent; receptacle flat, paleate; rays 4–8 per head, white, about 2 mm long, pistillate, fertile; disc flowers numerous, greenish, about 1.5 mm long, tubular, perfect, fertile, the corolla 4-lobed; cypselae of ray flowers 3- or 4-angled, usually glabrous, 2.0–2.5 mm long, of disc flowers flat; pappus a crown of short teeth, or absent.

Common Name: Yerba de tajo.
Habitat: Muddy shores, along streams, wet ditches, moist disturbed areas.
Range: New York and Ontario to South Dakota, south to Texas and Florida; New Mexico, Arizona, California; Mexico; Central America; West Indies; South America.
Illinois Distribution: Common in the southern three-fourths of Illinois, less common in the northern one-fourth where it may be adventive.

55. *Eclipta prostrata*
    (Yerba de tajo).
a. Upper part of plant.

b. Leaf.
c. Phyllary.
d. Flowering head.

e. Palea.
f. Disc flower.
g. Cypsela.

This species is readily recognized by its small flowering heads consisting of a few short, white rays around a flat, greenish disc, its opposite leaves, and its mauve-colored stem with white strigose hairs.

    I have been fascinated by its small flowering heads, which almost overnight become transformed into fruiting heads that consist of a flat, green disc subtended by 12–18 radiating, pointed, green bracts that appear star-like.

The nomenclature for this species has been confusing. Linnaeus recognized and named three types: a prostrate type he called *Verbesina prostrata* in 1753; a spreading type he named *Verbesina alba* in 1753; and an erect type he called *Eclipta erecta* in 1771. For many years, this plant has been called *Eclipta alba*.

Eclipta prostrata flowers from July to October.

50. **Acmella** Rich. in Pers.—Spilanthes

Annual or perennial herbs; stems prostrate to ascending to erect, branched or unbranched; leaves opposite, simple, petiolate; head radiate (in our species) or discoid, borne singly, pedunculate; involucre campanulate to ovoid; phyllaries up to 15, in 1 or 2 series, more or less equal, loosely appressed; receptacle conic, paleate; ray flowers up to 20, rarely absent, yellow (in Illinois), pistillate, fertile; disc flowers numerous, yellow, tubular, perfect, the corolla 4- or 5-lobed; cypselae of ray flowers 3-angled, of disc flowers flat and narrowly winged; pappus of 1–3 awns, or absent.

There are about thirty species in the genus, all native to the New World, with two in the United States.

For many years the species were placed in the genus *Spilanthes*.

1. **Acmella repens** (Walt.) Rich. in Pers. Syn. Pl. 2:473. 1807. Fig. 56.
Anthemis repens Walt. Fl. Carol. 211. 1788.
Spilanthes repens (Walt.) Michx. Fl. Bor. Am. 2:131. 1803.
Spilanthes americana (Mutis) Hieronymus var. *repens* (Walt.) A.H. Moore, Proc. Am.
    Acad. 42:547. 1907.
Acmella oppositifolia (Lam.) R.K. Jansen, Syst. Bot. Monogr. 8:34. 1985.

Perennial herb from slender rhizomes; stems spreading to ascending, rooting at the lower nodes, branched or unbranched, glabrous or pubescent, to 75 cm long; leaves opposite, simple, lanceolate to ovate, acute to acuminate at the apex, rounded or more or less truncate at the base, to 8 cm long, to 4 cm wide, coarsely serrate, glabrous or pubescent, on petioles to 3 cm long; head usually radiate, borne singly, on a long, glabrous peduncle; involucre campanulate to oblong; phyllaries 8–12 (–14), in 1–3 series, oblong to oblong-lanceolate, subequal; receptacle conic, paleate; ray flowers up to 20, rarely absent, yellow, 8–14 mm long, 3-notched at the apex, pistillate, fertile; disc flowers numerous, yellow, tubular, perfect, fertile, the corolla 4- or 5-lobed; cypselae oblong, hispidulous, ciliolate, 1.0–2.5 mm long, 3-angled from the ray flowers, flattened from the disc flowers; pappus of 1–3 short awns, or absent.

Common Name: Spilanthes; spot-flower.
Habitat: Woods.
Range: Illinois and Missouri, south to Texas and Florida.
Illinois Distribution: Known only from Johnson County.

This species is recognized by its flowering heads that have a short-conic disc surrounded by yellow rays 3-notched at the apex, and opposite leaves with coarse teeth.

In the past, this species has been known as *Spilanthes americana*.

Ernest J. Palmer made a train trip across southern Illinois in 1927. On that trip he reported seeing this species in Johnson County, but no specimens were apparently collected. In 2009, while hiking along an abandoned railroad near Tunnel Hill (Johnson County), I rediscovered this attractive species at the edge of a woods.

Acmella repens flowers in May and June.

56. *Acmella repens*
   (Spot-flower).
a. Habit.

b. Cluster of phyllaries.
c. Phyllary.
d. Flowering head.

e. Disc flower with palea.
f, g. Ray flowers.
h. Cypsela.

51. **Calyptocarpus** Less.—Carpet Daisy

Perennial herbs from slender rhizomes; stems prostrate, branched; leaves opposite, simple, petiolate; head radiate, borne singly in the axils of the leaves; involucre obconic; phyllaries 5 in 1 series, persistent; receptacle convex, paleate; ray flowers up to 8, yellow, pistillate, fertile; disc flowers up to 20, yellow, tubular, perfect, fertile, the corolla 4- or 5-lobed cypselae flat, tuberculate; pappus of 2–5 rather stout awns.

Two species comprise this genus in the United States, Mexico, and Central America.

This genus differs from the similar *Acmella* by its fewer disc flowers, its convex rather than conic receptacle, and its 2–5 stout awns.

1. **Calyptocarpus vialis** Less. Syn. Gen. Compos. 221. 1832. Fig. 57.

Perennial herb from slender rhizomes; stems much branched, prostrate, glabrous or somewhat pubescent, to 15 cm long; leaves opposite, simple, lanceolate to ovate, acute at the apex, truncate to tapering to the base, to 3.5 cm long, to 2.5 cm wide, serrulate, 3-nerved, pubescent and scabrous on both surfaces; head radiate, borne singly from the axils of the leaves; involucre obconic, to 8 mm across; phyllaries 5, in 1 series, to 7 mm long, persistent; receptacle convex, paleate, the paleae persistent, with a scarious margin; ray flowers up to 8, yellow, 2–5 mm long, pistillate, fertile; disc flowers up to 20, yellow, tubular, perfect, fertile, the corolla 4- or 5-lobed; cypselae oblong, flat, 2–4 mm long, tuberculate; pappus of 2 rather stout awns to 3 mm long.

Common Name: Carpet daisy.
Habitat: Disturbed soil.
Range: Texas to Florida; Mexico; Central America; adventive in Illinois.
Illinois Distribution: Known only from DeKalb County.

This species in Illinois, well north of its native range in the extreme southern United States, has been found as a persistent adventive in DeKalb County.

It is easily distinguished by its small, radiate, yellow flowering head from the leaf axils, its prostrate habit, and its opposite leaves.

Calyptocarpus vialis flowers during July and August.

Subtribe Helianthineae Cass ex Dum.

Annuals or perennials; leaves usually cauline, often opposite, entire or toothed; heads with ray flowers and disc flowers, single or in corymbs, panicles, or racemes; involucre cylindric to hemispheric; phyllaries in 1–6 series, unequal or subequal; receptacle flat or convex, paleate; ray flowers up to 100, usually much fewer, yellow or orange, neutral, sterile; disc flowers numerous, bisexual, yellow or orange, red or purple, 5-lobed; cypselae usually flattened, biconvex, obovoid to columnar to prismatic, glabrous or pubescent; pappus none or up to eight scales and/or awns.

Worldwide there are 17 genera and about 365 species in this subtribe. In Illinois, this subtribe includes only the genus *Helianthus*.

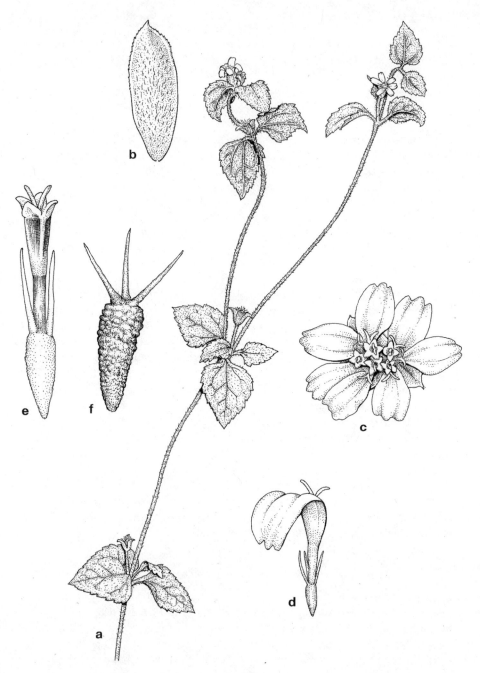

57. *Calyptocarpus vialis* (Carpet daisy).

a. Upper part of plant.
b. Phyllary.
c. Flowering head.

d. Ray flower.
e. Disc flower.
f. Cypsela.

52. **Helianthus** L.—Sunflower

Annual or perennial herbs, erect, usually branched, with opposite and/or alternate cauline leaves; heads solitary or in corymbs, with both ray flowers and disc flowers; involucre usually hemispheric; phyllaries in 2–3 series, unequal or subequal; receptacle flat or convex or low-conic, paleate; ray flowers up to 30 or more per head, yellow or orange, neutral; disc flowers numerous, yellow or reddish, bisexual, fertile, the corolla 5-lobed; cypselae obpyramidal, more or less flat, glabrous or pubescent; pappus of 2–3 large scales and several smaller scales.

There are about fifty species of *Helianthus*, all native to North America and Mexico. Nineteen species and six hybrids occur in Illinois.

1. Leaves with conspicuous ciliate margins; rays less than 1 cm long .... 1. *H. ciliaris*
1. Leaves without conspicuous ciliate margins; rays normally more than 1 cm long.
  2. Disc red or purple.
    3. Leaves linear, or rhombic, broadest near the middle.
      4. Stems and leaves glabrous, usually glaucous .... 2. *H. salicifolius*
      4. Stems and leaves strigose, pilose, or hispid, usually not glaucous.
        5. Stems and leaves strigose or pilose .... 3. *H. angustifolius*
        5. Stems and leaves hispid ....4. *H. subrhomboideus*
    3. Leaves lanceolate to ovate, broadest below the middle.
      6. Receptacle flat or nearly so.
        7. Phyllaries ovate, abruptly contracted above; receptacular bracts not bearded at apex. . . . . 5. *H. annuus*
        7. Phyllaries lanceolate, tapering to the tip; receptacular bracts bearded at apex. . . . . 6. *H. petiolaris*
      6. Receptacle convex or low-conic.
        8. Leaves tapering to short, thick petioles, or sessile .... 7. *H. pauciflorus*
        8. Leaves abruptly slender, petiolate....................... 8. *H. silphioides*
  2. Disc yellow.
    9. Stems scapose, or with 3–5 pairs of cauline leaves smaller than the basal leaves . . . . 9. *H. occidentalis*
    9. Stems leafy to the inflorescence.
      10. Stems glabrous or nearly so below the inflorescence, often glaucous.
        11. Heads 1.5–3.0 cm across; disc 0.4–1.0 cm across; rays 1.0–1.5 cm long . . . . 10. *H. microcephalus*
        11. Heads 4–9 cm across; disc 1.0–2.5 cm across; rays 2–4 cm long.
          12. Leaves thin, membranous ....11. *H. decapetalus*
          12. Leaves thick, firm.
            13. Most or all the leaves alternate .... 12. *H. grosseserratus*
            13. Most leaves opposite, except sometimes the uppermost.
              14. Leaves sessile or on petioles up to 5 mm long .. 13. *H. divaricatus*
              14. Leaves petiolate, the petioles more than 5 mm long . . . . 14. *H. strumosus*
      10. Stems pubescent or scabrous throughout, rarely glaucous.
        15. Leaves gray-pubescent on both surfaces.
          16. Leaves densely gray-pubescent on both surfaces; cypselae villous, at least near tip; leaves lance-ovate to lanceolate, up to 18 cm long.

17. Ray flowers 20–22 per head; heads 2.0–2.5 cm across; leaves lance-ovate to lanceolate .... 15. *H. mollis*
17. Ray flowers 17–19 per head; heads 1.5–2.0 cm across; leaves lanceolate.................................... 16. *H. X cinereus*
16. Leaves sparsely gray-pubescent on both surfaces; cypselae usually glabrous; leaves lanceolate, over 20 cm long ....17. *H. X brevifolius*
15. Leaves not gray-pubescent.
   18. Petioles 2–8 cm long.
      19. Phyllaries acute at tip........................... 18. *H. X laetiflorus*
      19. Phyllaries acuminate to long-attenuate at tip .... 19. *H. tuberosus*
   18. Petioles up to 2 cm long, or absent.
      20. All leaves opposite except for the upper 2 or 3.
         21. Main side vein of leaves joining midvein at base of blade; leaves broadest at base............................. 20. *H. hirsutus*
         21. Main side vein of leaves joining midvein about 1 cm above base of blade; leaves broadest near middle.......... 21. *H. X doronicoides*
      20. Leaves opposite on the lower half of the stem, alternate on the upper half.
         22. Stems with spreading hairs................... 22. *H. giganteus*
         22. Stems with appressed hairs or short straight hairs or merely scabrous.
            23. Stems with appressed hairs or short straight hairs.
               24. Stems with short straight hairs...... 23. *H. X intermedius*
               24. Stems with appressed hairs .... 24. *H. maximilianii*
            23. Stems merely scabrous .... 25. *H. X luxurians*

1. **Helianthus ciliaris** DC. Prodr. 5:587. 1836. Fig. 58.

Perennial herbs with rhizomes, sometimes forming colonies; stems usually erect, glabrous, usually glaucous, to 75 cm tall; leaves usually opposite, usually blue-green, linear to lanceolate, acute at the apex, tapering to the base, 5.0–7.5 cm long, 0.5–2.0 cm wide, sessile, entire or serrate, glabrous to hispid, with ciliate margins; heads radiate, 1–5, on peduncles up to 12 cm long; involucre hemispheric, 12–25 mm across; phyllaries 16–20, ovate or narrowly ovate, 3–8 mm long, 2.0–3.5 mm wide, obtuse to acute at the apex, glabrous or strigose, eglandular, with ciliate margins; receptacle low-conic, paleate, the paleae about 7 mm long, pubescent, glandular; ray flowers 10–18, yellow, 8–9 mm long, neutral; disc flowers 38–50, reddish, tubular, bisexual, fertile, the corolla 5-lobed, 4–6 mm long; cypselae flattened, 3.0–3.5 mm long, glabrous; pappus of two aristate scales 1.2–1.5 mm long.

Common Name: Texas blueweed; ciliate sunflower.
Habitat: Disturbed soil (in Illinois).
Range: Nebraska to California, east to Texas; adventive elsewhere in the United States.
Illinois Distribution: Adventive in St. Clair County.

This species of the western United States has been found a single time in Illinois in disturbed soil. It differs from other species of *Helianthus* by its distinctly ciliate

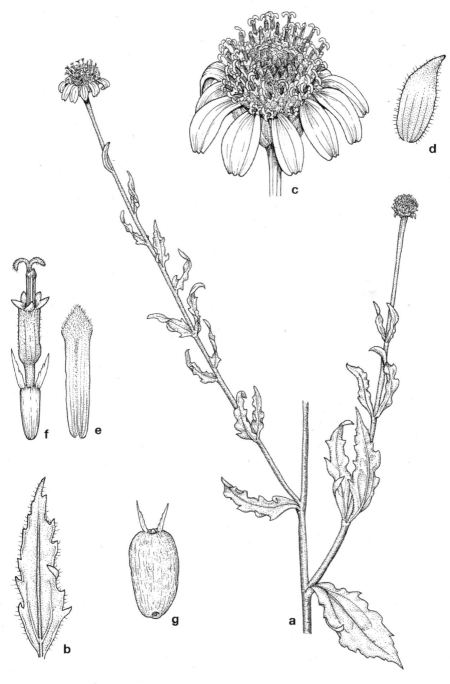

58. *Helianthus ciliaris*
    (Texas blueweed).
a. Upper part of plant.

b. Leaf.
c. Flowering head.
d. Phyllary.

e. Palea.
f. Disc flower.
g. Cypsela.

leaf margins, its usually blue-green leaves, and its short rays that are less than 1 cm long.

This species flowers in August and September.

2. Helianthus salicifolius A. Dietr. Allg. Gartenz. 2:337. 1834. Fig. 59.

Perennial herbs with rhizomes; stems erect, glabrous, usually glaucous, to 2.5 cm tall; leaves numerous, crowded, alternate, linear to narrowly lanceolate, acute

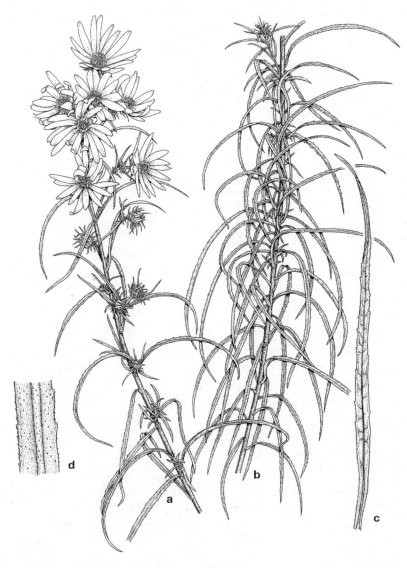

59. *Helianthus salicifolius*
(Willow-leaved
sunflower).

a. Upper part of plant.
b. Section of stem and
leaves.

c. Leaf.
d. Portion of leaf.

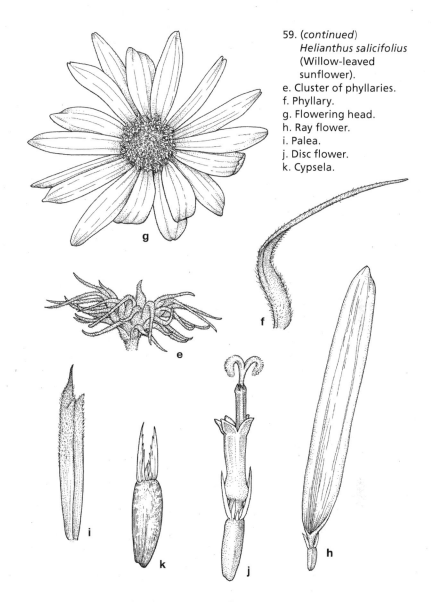

59. (*continued*)
    Helianthus salicifolius
    (Willow-leaved
    sunflower).
e. Cluster of phyllaries.
f. Phyllary.
g. Flowering head.
h. Ray flower.
i. Palea.
j. Disc flower.
k. Cypsela.

at the apex, tapering to the base, to 20 mm long, 2–15 mm wide, sessile, serrulate to entire, scabrous, glandular; heads radiate, up to 15, in panicles, on peduncles up to 6 cm long; involucre campanulate, up to 18 mm across; phyllaries up to 50, spreading, lance-linear, to 20 mm long, to 2 mm wide, subequal, more or less glabrous, ciliate, eglandular; receptacle paleate, the paleae 8–10 mm long; ray flowers up to 20 per head, yellow, 2.5–3.5 cm long, neutral; disc flowers about 50, reddish, tubular, bisexual, fertile, the corolla 5-lobed, 5–6 mm long, the disc up to 1 cm across; cypselae obovoid, flattened, 4–6 mm long, glabrous; pappus of 2 aristate scales 3.0–3.5 mm long, plus up to 8 tiny scales.

Common Name: Willow-leaved sunflower.
Habitat: Disturbed soil (in Illinois).
Range: Missouri to Nebraska, south to Texas; adventive elsewhere.
Illinois Distribution: Adventive in Cook County, but apparently now extirpated.

This is a species of the west-central United States. The only collection was discovered in disturbed soil.

This is the only species of *Helianthus* in Illinois with a reddish disc and linear glabrous and glaucous leaves.

Helianthus salicifolius flowers from August to October.

3. **Helianthus angustifolius** L. Sp. Pl. 2:906. 1753. Fig. 60.
Perennial herbs with a caudex but no rhizomes; stems erect, branched, strigose or pilose or often scabrous, up to 1.5 m tall; leaves opposite or sometimes mostly alternate, simple, firm, linear to linear-lanceolate, often in axillary fascicles, acute to acuminate at the apex, tapering to the base, to 15 (–20) cm long, 3–15 mm wide, sessile, sometimes revolute, entire, strigose, pilose, or hispidulous, scabrous, usually glandular; heads radiate, 3–15, rarely solitary, on peduncles up to 15 cm long; involucre hemispheric, 10–20 mm across; phyllaries up to 30, loosely arranged, lanceolate, to 10 mm long, 1–2 mm wide, scabrous, usually glandular; receptacle paleate, the paleae 5.5–6.5 mm long; rays up to 20 per head, yellow, erose at the apex, 12–20 mm long, neutral; disc flowers numerous, red-purple, tubular, bisexual, fertile, the corolla 5-lobed, 4.0–4.5 mm long, the disc up to 1 cm across; cypselae obovoid, flattened, 2–4 mm long, glabrous or nearly so; pappus of 2 aristate scales 1.5–2.0 mm long.

Common Name: Narrow-leaved sunflower; swamp sunflower.
Habitat: Moist ground, sometimes in wet ditches.
Range: Pennsylvania to Iowa, south to Texas and Florida.
Illinois Distribution: Known from Massac, Pope, Pulaski, and Wayne counties in the southern one-sixth of the state.

This species is distinguished by its red-purple disc and its linear to linear-lanceolate firm, scabrous leaves.

Helianthus angustifolius is found in wet ground in the southern one-sixth of the state.

This species flowers from August to October.

4. **Helianthus subrhomboideus** Rydb. Mem. N.Y. Bot. Gard. 1:419. 1900. Fig. 61.
Helianthus laetiflorus Pers. var. *subrhomboideus* (Rydb.) Fern. Rhodora 48:79. 1946.
Helianthus rigidus (Cass.) Desf. ssp. *subrhomboideus* (Rydb.) Heiser, Mem. Torrey Club 22:136. 1969.
Helianthus rigidus (Cass.) Desf. var. *subrhomboideus* (Rydb.) Cronq. Fl. Pacif. Northwest 528. 1973.
Helianthus pauciflorus Nutt. ssp. *subrhomboideus* (Rydb.) O. Spring & E.E. Schilling, Biochem. Syst. & Ecol. 18:22. 1990.

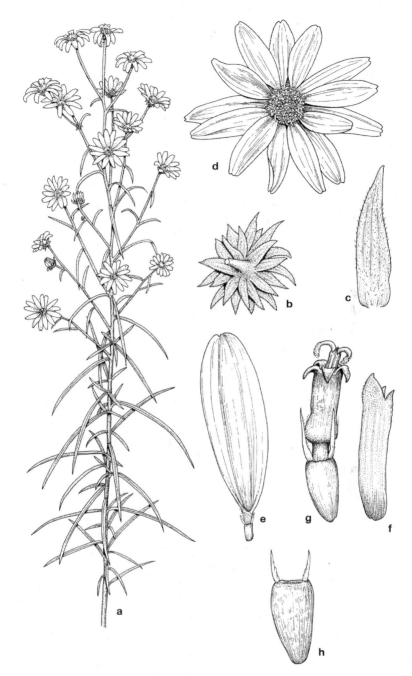

60. *Helianthus angustifolius* (Narrow-leaved sunflower).

a. Upper part of plant.
b. Cluster of phyllaries.
c. Phyllary.
d. Flowering head.

e. Ray flower.
f. Palea.
g. Disc flower.
h. Cypsela.

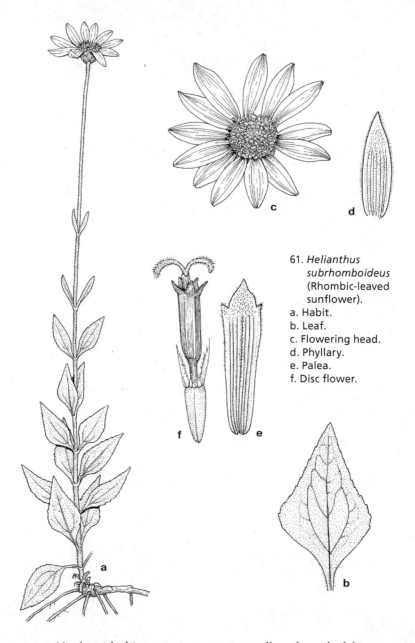

61. *Helianthus subrhomboideus* (Rhombic-leaved sunflower).
a. Habit.
b. Leaf.
c. Flowering head.
d. Phyllary.
e. Palea.
f. Disc flower.

Perennial herbs with rhizomes; stems erect, usually unbranched, hirsute to hispid, to 1.2 m tall; leaves opposite, simple, firm, rhombic-ovate to rhombic-lanceolate, obtuse to acute at the apex, tapering to the usually petiolate base, to 12 cm long, to 4 cm wide, the uppermost and lowermost smaller, hispid, scabrous, serrulate, usually glandular; heads radiate, often solitary or up to 3 on short-hirsute peduncles, the peduncles up to 10 cm long; involucre hemispheric, up to 20 mm across; phyllaries up to 30, oval or oblong, hispidulous, ciliate, to 10 mm long, to 5 mm wide; receptacle paleate, the paleae about 10 mm long; rays 12–20 per head, yellow, 2–3 cm long, neutral; disc flowers numerous, red-purple, tubular, bisexual, fertile, the corolla 5-lobed, 6–7 mm long, the disc up to 1.2 cm across; cypselae obovoid, flattened, 5–6 mm long, glabrous or nearly so; pappus of 2 aristate scales 4–5 mm long and a few smaller scales.

Common Name: Rhombic-leaved sunflower; prairie sunflower; stiff sunflower.
Habitat: Dry prairies.
Range: Nova Scotia to Alberta, south to New Mexico, Texas, Missouri, Illinois, and Pennsylvania.
Illinois Distribution: Scattered in the northern half of the state.

As noted in the synonymy, this species in the past has been considered a variation of *Helianthus rigidus* (Cass.) Desf., *H. laetiflorus* Pers., or *H. pauciflorus* Nutt. Wilhelm and Rericha (2017) consider this plant to be a subspecies of *H. pauciflorus*.

The red-purple disc and the rhombic-ovate or rhombic-lanceolate leaves are distinctive.

Helianthus subrhomboideus flowers from July to October.

5. **Helianthus annuus** L. Sp. Pl. 2:904. 1753. Fig. 62.
Helianthus lenticularis Dougl. ex Lindl. in Edwards, Bot. Reg. 15:plate 1265. 1829.

Robust annual with fibrous roots; stems erect, stout, hispid, to 4 m tall; lowest leaves sometime opposite, the others alternate, broadly ovate, abruptly acuminate at the apex, cordate or rounded at the petiolate base, to 40 cm long, 12–20 cm wide, 3-nerved, serrate to dentate, hispid, sometimes glandular; heads radiate, solitary or few, on hispid peduncles up to 20 cm long; involucre hemispheric, the 8 cm across; phyllaries 20 or more, ovate to lance-ovate, acute to usually abruptly long-acuminate at the apex, 12–25 mm long, to 7.5 mm wide, hispid, ciliate, glandular; receptacle flat, paleate, the paleae about 10 mm long; ray flowers numerous, yellow, up to 5 cm long, neutral; disc flowers numerous, usually reddish, tubular, bisexual, fertile, the corolla 5-lobed, up to 8 mm long, the disc up to 60 cm across; cypselae obovoid to suborbicular, 4–12 mm long, glabrous for the pilose apex; pappus of 2 lanceolate scales up to 3.5 mm long, sometimes with a few tiny scales in addition.

Common Name: Common sunflower.
Habitat: Fields, roadsides.
Range: Native west of Illinois; common as an adventive in Illinois.
Illinois Distribution: Common throughout the state.

This is a common sunflower found in fields and along roads throughout Illinois.

Helianthus annuus differs from the similar *H. petiolaris* by its wider leaves and broader phyllaries that are usually abruptly acuminate at the apex.

62. *Helianthus annuus* (Common sunflower).

a. Upper part of plant.

b. Leaf.
c. Cluster of phyllaries.
d. Phyllary.
e. Flowering head.

f. Ray flower.
g. Palea.
h. Disc flower.
i. Cypsela.

Specimens as small as ten centimeters high may flower, while other specimens may be several meters tall.

Helianthus annuus flowers from July to November.

6. **Helianthus petiolaris** Nutt. Journ. Acad. Nat. Sci. Phila. 2:115. 1821. Fig. 63.

Annual herbs with fibrous roots; stems erect, branched, hispid to strigose, to 2.5 m tall; lowest leaves sometimes opposite, the others alternate, lance-ovate, less commonly ovate, acute at the apex, tapering or rounded at the petiolate base, to 15 cm long, to 8 cm wide, 3-nerved, entire or dentate, scabrous, strigose, sometimes glandular; heads radiate, solitary or up to 4 or 5, on scabrous peduncles up to 20 cm long; involucre hemispheric, up to 2 cm across; phyllaries 15–25, narrowly lanceolate to lanceolate, tapering to the acuminate and bearded apex, 10–14 mm long, 3–5 mm wide, usually hispidulous; receptacle flat, paleate, the paleae up to 7 mm long; ray flowers up to 30 per head, yellow, up to 2 cm long, neutral; disc flowers numerous, usually reddish, tubular, bisexual, fertile, the corolla 5-lobed, 4–6 mm long, the disc up to 2.5 cm across; cypselae obovoid, 3.0–4.5mm long, pubescent; pappus of 2 aristate scales up to 3 mm long, sometimes with very tiny scales as well.

Common Name: Prairie sunflower.
Habitat: Sandy soil in fields, along roads, and along railroads.
Range: Native to the western United States; adventive in Illinois.
Illinois Distribution: Occasional throughout the state.

This species often looks like a smaller version of *Helianthus annuus*, differing by its narrower leaves and narrower and shorter phyllaries.

Helianthus petiolaris flowers from June to November.

7. **Helianthus pauciflorus** Nutt. Gen. N. Am. Pl. 2:177. 1818. Fig. 64.
Harpalium rigidum Cass. Bull. Sci. Nat. 20:200. 1814.
Helianthus scaberrimus Ell. Bot. S. C. & Ga. 2:423. 1824.
Helianthus rigidus (Cass.) Desf. Tabl. Ecole. Bot. ed. 3, 184. 1829.
Helianthus laetiflorus Pers. var. *rigidus* (Cass.) Fern. Rhodora 48:79. 1946.

Perennial herbs with rhizomes; stems erect, branched, hispid, to 2.0 (–2.5) m tall; leaves mostly basal, the cauline ones alternate, narrowly ovate, acute at the apex, to 20 cm long, to 6 cm wide, serrate or less commonly entire, usually hispid, sometimes glandular; heads radiate, 1–10, on strigose peduncles up to 12 cm long; involucre hemispheric, up to 20 mm across; phyllaries 25–35, ovate, acute at the apex, glabrous or hispidulous, ciliate, 6–10 mm long, 3–5 mm wide; receptacle convex to low-conic, paleate, the paleae about 10 mm long; ray flowers 10–20 per head, yellow, 2.0–3.5 cm long, neutral; disc flowers numerous, reddish, tubular, bisexual, fertile, the corolla 5-lobed, 6–7 mm long; cypselae obovoid, 5–6 mm long, glabrous or nearly so; pappus of 2 aristate scales 4–5 mm long, with a few tiny scales as well.

63. *Helianthus petiolaris*
(Prairie sunflower).
a, b. Upper part of
plant.

c. Leaf.
d. Cluster of phyllaries.
e, f. Phyllaries.
g. Flowering head.

h. Ray flower.
i. Palea.
j. Disc flower.
k. Cypsela.

64. *Helianthus pauciflorus* (Prairie sunflower).

a. Upper part of plant.
b. Leaf.
c. Cluster of phyllaries.
d. Phyllary.

e. Flowering head.
f. Ray flower.
g. Palea.
h. Disc flower.

Common Name: Prairie sunflower; stiff sunflower.
Habitat: Dry or mesic prairies.
Range: Quebec to British Columbia, south to New Mexico and Georgia.
Illinois Distribution: Occasional throughout Illinois.

This species is distinguished by its reddish disc, its convex to low-conic receptacle, and its narrowly ovate leaves tapering to thick, short petioles or sessile.

Helianthus pauciflorus flowers from July to October.

8. Helianthus silphioides Nutt. Trans. Am. Phil. Soc. n. s., 7:366. 1841. Fig. 65.
Silphium atrorubens L. var. *pubescens* Kuntze, Rev. Gen. 1:343. 1891.

Perennial herbs from rhizomes, often forming colonies; stems erect, branched, up to 3 m tall, hispid; leaves basal and cauline, the cauline opposite, ovate to suborbicular, obtuse to subacute at the apex, usually rounded at the base, to 15

65. *Helianthus silphioides* (Rosinweed sunflower).
a. Upper part of plant.
b. Stem with cauline leaves.
c. Cluster of phyllaries.
d. Phyllary.
e. Flowering head.
f. Ray flower.
g. Palea.
h. Disc flower.
i. Cypsela.

cm long, about as wide, serrate or entire, scabrous at least on the lower surface, eglandular, on slender petioles up to 5 cm long; heads radiate, few to about 15, on peduncles up to 10 cm long; involucre hemispheric, up to 20 mm across; phyllaries 16–20 (–23), oblong to obovate, obtuse at the apex, 8–10 mm long, 3–5 mm wide, more or less glabrous, ciliate; receptacle convex to low-conic, paleate, the paleae about 10 mm long; ray flowers 8–12 per head, yellow, 15–20 mm long, neutral; disc flowers numerous, reddish, tubular, bisexual, fertile, the corolla 5-lobed, 6–7 mm long; cypselae oblongoid, 3–4 mm long, pubescent near the apex, more or less glabrous near the base; pappus of 2 aristate scales about 2.5 mm long.

Common Name: Rosinweed sunflower.
Habitat: Prairies.
Range: Illinois to Oklahoma, south to Louisiana and Alabama.
Illinois Distribution: Very rare; Alexander and St. Clair counties.

This species resembles *Silphium integrifolium* because of its opposite cauline leaves. It differs by its reddish disc and its smaller oblongoid cypselae.

This species may be extirpated from Illinois. It flowers from July to October.

9. **Helianthus occidentalis** Riddell, W.J. Med. Phys. Sci. 9:577. 1836. Fig. 66.
Helianthus dowellianus M.A. Curtis, Am Journ. Sci. & Arts 44:82. 1843.
Helianthus illinoensis Gleason, Ohio Nat. 5:214. 1904.
Helianthus occidentalis Riddell var. *dowellianus* (M.A. Curtis) Torr. & Gray, Fl. N. Am.
2:504. 1843.
Helianthus occidentalis Riddell var. *illinoensis* (Gleason) Gates, Bull. Torrey Club
37:81. 1910.

Perennial herbs with rhizomes and often with stolons; stems often unbranched, slender, usually scapose or nearly so, erect, to 1.5 m tall, pilose to hirsute at the base, usually glabrous or nearly so above; leaves mostly basal or crowded near the base and opposite, elliptic to narrowly ovate, acute at the apex, tapering to the petiolate base, to 15 cm long, to 7 cm wide, scabrous and hispidulous, usually entire, glandular, the upper leaves, if present, often in 3–5 pairs, reduced and sometimes bract-like; heads radiate, solitary or up to 6 (–10), on usually glabrous peduncles up to 12 cm long; involucre cylindric, 10–15 mm across; phyllaries 20–25, appressed, lanceolate to narrowly ovate, acuminate at the apex, very unequal, 4–10 mm long, 1.5–2.5 mm wide, usually glabrous, ciliate; receptacle flat, paleate, the paleae 5–7 mm long; ray flowers 10–15 per head, yellow, 1–2 cm long, neutral; disc flowers 10–15, yellow, tubular, bisexual, fertile, the corolla 5-lobed, 4.5–5.5 mm long; cypselae oblongoid, 3.0–4.5 mm long, usually pubescent at the apex; pappus of 2 aristate scales 1.5–3.0 mm long, sometimes with a few tiny scales.

Common Name: Western sunflower; fewleaf sunflower.
Habitat: Sand prairies, hill prairies, black oak savannas.
Range: Massachusetts to Wisconsin, south to Kansas, Arkansas, and Florida.

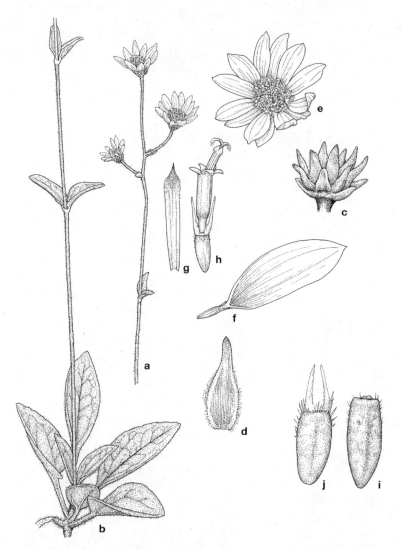

66. *Helianthus*
  occidentalis (Western
  sunflower).
a. Upper part of plant.

b. Lower part of plant.
c. Cluster of phyllaries.
d. Phyllary.
e. Flowering head.

f. Ray flower.
g. Palea.
h. Disc flower.
i, j. Cypselae.

Illinois Distribution: Scattered in the northern half of the state, extending south to
  St. Clair County.

This species is readily distinguished by its scapose or nearly scapose growth form.
    The type specimen for *H. illinoensis* is from sand dunes along the Illinois River
near Havana, in Mason County, collected on August 17, 1904.
    Helianthus occidentalis is a species of prairies. It flowers from July to October.

10. **Helianthus microcephalus** Torr. & Gray, Fl. N. Am. 2:329. 1842. Fig. 67.

Perennial herbs from a slender rhizome; stems erect, branched, to 2 m tall, glabrous; leaves usually opposite, a few sometimes alternate above, thin, lanceolate, acute at the apex, tapering to the petiolate base, to 10 cm long, to 3 cm wide, 3-nerved, entire or serrate, usually tomentose on the lower surface, glandular; heads radiate, up to 15, on glabrous peduncles up to 8 cm long; involucre campanulate, 5–7 mm across; phyllaries 12–17, narrowly lanceolate, acuminate at the apex, to 6.5 mm long, 1.5–2.5 mm wide, glabrous except for the ciliate margin;

67. *Helianthus microcephalus* (Small woodland sunflower).
a. Upper part of plant.
b. Leaf.
c. Cluster of phyllaries.
d. Phyllary.
e. Flowering head.
f. Ray flower.
g. Disc flower with subtending palea.
h. Cypsela.

receptacle flat, paleate, the paleae 5–7 mm long; ray flowers up to 8, yellow, 10–15 mm long, neutral; disc flowers 15–20, yellow, tubular, bisexual, perfect, the corolla 5-lobed, 4–5 mm long, the disc up to 6 mm across; cypselae oblongoid, 3.5–4.0 mm long, glabrous; pappus of 2 aristate scales 1.5–2.0 mm long.

Common Name: Small woodland sunflower.
Habitat: Dry open woods.
Range: Connecticut to Minnesota, south to Louisiana and Florida.
Illinois Distribution: Occasional in southern Illinois.

The small woodland sunflower is readily recognized by its smaller flowering heads with up to 8 rays and about 20 disc flowers. The stems are glabrous. The leaves are usually tomentose on the lower surface.

   Helianthus microcephalus flowers from August to October.

   11. **Helianthus decapetalus** L. Sp. Pl. 2:905. 1753. Fig. 68.
Helianthus frondosus L. Amoen. Acad. 4:290. 1759.
Helianthus decapetalus L. var. *frondosus* (L.) Gray, Man. Bot. 227. 1848.

   Perennial herbs with rhizomes; stems erect, branched, to 2 m tall, glabrous; leaves opposite except sometimes for the uppermost, thin, ovate-lanceolate to ovate, acute to acuminate at the apex, rounded or tapering to the petiolate base and decurrent part way on the petiole, 3-nerved, to 17 cm long, 4–8 cm wide, scabrous-hispid, serrate, sparsely glandular, the petioles up to 5 mm long; heads radiate, 3–10, on scabrous peduncles up to 12 cm long; involucre hemispheric, 12–25 mm across; phyllaries 12–25, spreading to reflexed, narrowly lanceolate, 10–15 mm long, acuminate at the apex, glabrous to strigillose, ciliate; receptacle flat, paleate, the paleae 8–10 mm long; ray flowers 8–15 per head, pale yellow, 2.0–2.5 cm long, neutral; disc flowers 40 or more, yellow, tubular, bisexual, fertile, the corolla 5-lobed, 6.5–7.0 mm long, the disc 1.0–1.5 cm across; cypselae obovoid, 3.5–5.0 mm long, glabrous; pappus of 2 aristate scales 3–4 mm long.

Common Name: Thinleaf sunflower; pale sunflower.
Habitat: Dry or mesic woods, savannas.
Range: New Brunswick to Wisconsin, south to Oklahoma, Louisiana, and Georgia.
Illinois Distribution: Occasional throughout the state.

The pale rays, thin leaves, and glabrous stems are the distinguishing characteristics for this species.

   Helianthus decapetalus flowers from July to October.

   12. **Helianthus grosseserratus** M. Martens, Index Sem. (Louvain) 1839. Figs. 69a (a–h), 69b (i, j).
Helianthus instabilis E.E. Wats. Papers Mich. Acad. Sci. 9:423. 1929.
Helianthus grosseserratus M. Martens f. *pleniflorus* Wadmond, Rhodora 3419. 1932.

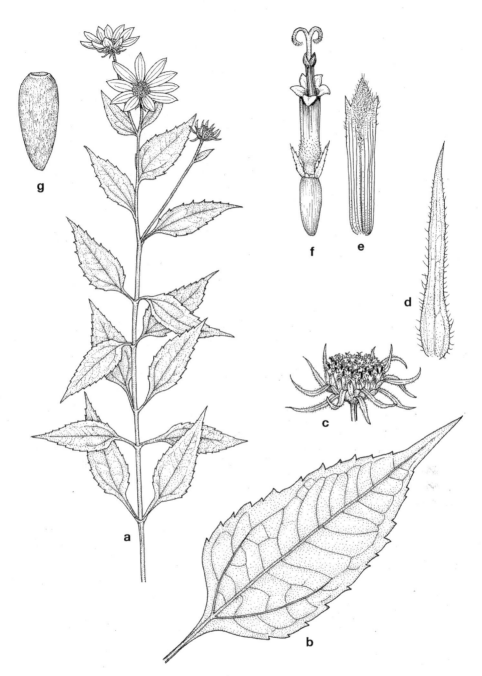

68. *Helianthus decapetalus* (Thinleaf sunflower).

a. Upper part of plant.
b. Leaf.
c. Flowering head with rays removed.
d. Phyllary.
e. Palea.
f. Disc flower.
g. Cypsela.

69a. *Helianthus grosseserratus* (Sawtooth sunflower).

a. Upper part of plant.
b. Leaf.
c. Phyllary.
d. Flowering head.
e. Ray flower.
f. Palea.
g. Disc flower.
h. Cypsela.

Robust perennial herbs from rhizomes; stems erect, branched, to 4 m tall, glabrous, often glaucous, rarely strigillose; leaves usually opposite below, alternate above, lanceolate to narrowly ovate, acuminate at the apex, tapering to the petiolate base, to 30 cm long, 4–8 cm wide, 3-nerved, serrate, rarely entire, scabrous above, densely pubescent beneath, glandular; heads radiate, up to 15 (–20), on usually glabrous peduncles up to 10 cm long; involucre hemispheric, to 25 mm across; phyllaries up to 30, linear-lanceolate, spreading or even squarrose, glabrous or puberulent, ciliate, 10–12 mm long, 1.5–2.5 mm wide; receptacle flat, paleate, the paleae 7–8 mm long; rays radiate, 10–20 per head, yellow, 2.5–4.0 cm long, neutral; disc flowers numerous, yellow, tubular, bisexual, fertile, the corolla 5-lobed, 5–6 mm long, the disc 10–25 mm across; cypselae oblongoid, 3–4 mm long, glabrous or nearly so; pappus of 2 aristate scales 2.0–2.5 mm long.

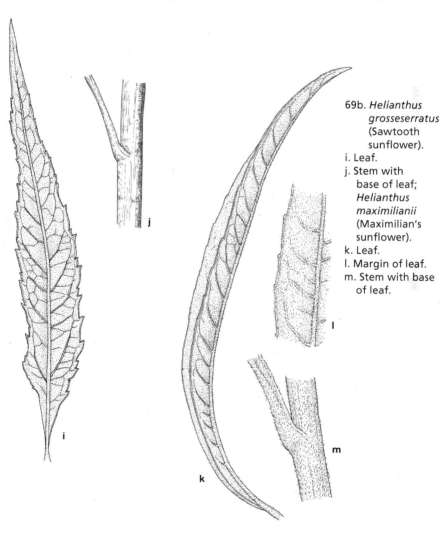

69b. *Helianthus grosseserratus* (Sawtooth sunflower).
i. Leaf.
j. Stem with base of leaf; *Helianthus maximilianii* (Maximilian's sunflower).
k. Leaf.
l. Margin of leaf.
m. Stem with base of leaf.

Common Name: Sawtooth sunflower.
Habitat: Prairies, edge of woods, wet roadsides, wet ditches, sedge meadows, fens, marshes.
Range: Quebec to North Dakota, south to Texas and Georgia.
Illinois Distribution: Common throughout the state.

Although most specimens have leaves that are serrate, a few specimens have leaves with the margins entire or nearly so. Most stems are glabrous and often glaucous, although a few specimens have strigillose stems.

Helianthus grosseserratus resembles *H. giganteus*, but differs readily by its glabrous or rarely strigillose stems.

The type specimen for *H. instabilis* E.E. Wats. is from Casey in Clark County, but I consider it synonymous with *H. grosseserratus*.

This species flowers from June to October.

13. **Helianthus divaricatus** L. Sp. Pl. 2:906. 1753. Fig. 70.

Perennial herbs from rhizomes; stems erect, branched or unbranched, to 2 m tall, glabrous or slightly pubescent, sometimes glaucous; leaves opposite, or the uppermost rarely alternate, thick, horizontally spreading, lanceolate to lance-ovate, acute to acuminate at the apex, rounded or subcordate at the sessile or very short-petiolate base, to 12 cm long, to 4.5 cm wide, 3-nerved, usually serrate, scabrous above, hispid beneath; heads radiate, up to 10 per head, on scabrous peduncles up to 7 cm long; involucre hemispheric, 10–15 mm across; phyllaries up to 25, somewhat recurved, linear-lanceolate to narrowly ovate, acuminate at the apex, to 12 mm long, 2.0–2.5 mm wide, hispidulous to nearly glabrous, ciliate; receptacle flat, paleate, the paleae up to 8 mm long; ray flowers 8–12, yellow, 1.5–3.0 cm long, neutral; disc flowers 40–50, yellow, tubular, bisexual, fertile, the corolla 5-lobed, 4.0–5.5 mm long, the disc 10–12 mm across; cypselae oblongoid, 3.0–3.5 mm long, glabrous or nearly so; pappus of 2 aristate scales up to 2.5 mm long.

Common Name: Woodland sunflower.
Habitat: Open woods, black oak savannas, rocky glades, prairies.
Range: Maine to Wisconsin, south to Oklahoma and Florida.
Illinois Distribution: Common throughout the state.

Helianthus divaricatus differs from *H. hirsutus* by its glabrous stems and from *H. strumosus* by its sessile or nearly sessile leaves.

This species flowers from July to October.

14. **Helianthus strumosus** L. Sp. Pl. 2:905. 1753. Fig. 71.

Helianthus strumosus L. var. *mollis* (Willd.) Torr. & Gray, Fl. N. Am. 2:327. 1843.
Helianthus formosus E.E. Wats. Papers Mich. Acad. Sci. 9:445. 1929.
Helianthus leoninus E.E. Wats. Papers Mich. Acad. Sci. 9:449. 1929.
Helianthus arenicola E.E. Wats. Papers Mich. Acad. Sci. 9:453. 1929.

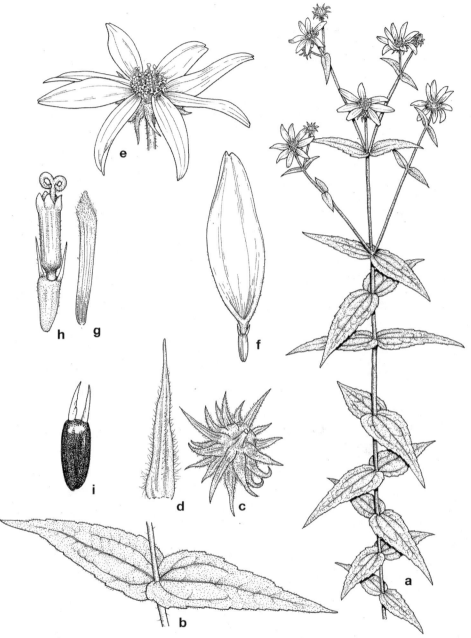

70. *Helianthus divaricatus* (Woodland sunflower).

a. Upper part of plant.

b. Node with opposite leaves.
c. Cluster of phyllaries.
d. Phyllary.
e. Flowering head.

f. Ray flower.
g. Palea.
h. Disc flower.
i. Cypsela.

71. *Helianthus strumosus* (Paleleaf woodland sunflower).
a. Upper part of plant.
b. Node with opposite leaves.
c. Rootstock.
d. Cluster of phyllaries.
e. Phyllary.
f. Flowering head.
g. Ray flower.
h. Palea.
i. Disc flower.
j. Cypsela.

Perennial herbs from rhizomes; stems erect, branched, to 3 m tall, glabrous or nearly so; leaves opposite except sometimes for the uppermost, thick, lance-ovate to ovate, acute to acuminate at the apex, tapering or subcordate at the petiolate base, to 15 cm long, to 8 cm wide, 3-nerved, serrate or occasionally entire, scabrous above, pubescent beneath, glandular; heads radiate, up to 15, on usually slightly pubescent peduncles up to 8 cm long; involucre cylindric to hemispheric, to 20 mm across; phyllaries 18–25, usually spreading or squarrose, sometimes erect, lanceolate, 5–10 mm long, 1.5–3.0 mm wide; receptacle flat, paleate, the paleae 5.0–6.5 mm long; ray flowers 10–18, yellow, to 2.0 (–2.5) cm long, neutral; disc flowers 35–50, yellow, tubular, bisexual, fertile, the corolla 5-lobed, 5.5–6.5 mm long; cypselae oblongoid, 4.0–5.5 mm long, glabrous; pappus of 2 aristate scales 2.0–2.5 mm long.

Common Name: Paleleaf woodland sunflower.
Habitat: Open woods, savannas, prairies.
Range: Maine to North Dakota, south to Texas and Florida.
Illinois Distribution: Occasional to common throughout the state.

Helianthus strumosus is similar to *H. divaricatus* because of the firm leaves and glabrous stems, but differs by its petioles up to 3 cm long.

This species flowers from July to October.

15. **Helianthus mollis** Lam. Encycl. 3:85. 1789. Figs. 72a, 72b.
Helianthus pubescens Vahl, Symb. Bot. 29:92. 1791.
Helianthus canescens Michx. Fl. Bor. Am. 2:140. 1803.

Perennial herbs from rhizomes; stems erect, often unbranched, to 1.2 m tall, hirsute or villous; leaves opposite except for sometimes the uppermost, simple, lance-ovate to broadly ovate, acute to acuminate at the apex, rounded or cordate at the sessile or sometimes clasping base, to 18 cm long, to 6 cm wide, gray-green, 3-nerved, serrulate or sometimes entire, gray-hispid to gray-tomentose, glandular; heads radiate, solitary or up to 12 (–15), on scabrous peduncles up to 15 cm long; involucre hemispheric, up to 2.5 cm across; phyllaries 30–40, lanceolate, 10–15 mm long, 2.0–3.5 mm wide, densely pubescent; receptacle flat, paleate, the paleae about 10 mm long; ray flowers 20–22, yellow, 2.5–3.0 cm long, neutral; disc flowers numerous, yellow, tubular, bisexual, fertile, the corolla 5-lobed, 6.0–7.5 mm long, the disc 2–3 cm across; cypselae oblongoid, 3.5–4.0 mm long, glabrous or sometimes pubescent at the apex; pappus of 2 aristate scales about 3 mm long.

Helianthus mollis has very showy flowering heads. It is distinguished by its hirsute to villous stems and its gray-pubescent leaves.

This species forms a hybrid with *H. occidentalis* called *H. X cinereus*. It also forms a hybrid with *H. giganteus* called *H. X doronicoides*. It forms a hybrid with *H. grosseserratus* called *H. X brevifolius*.

Two varieties occur in Illinois:
a. Leaves rounded at the sessile base . . . . .15a. *H. mollis* var. *mollis*
a. Leaves cordate at the clasping base . . . . .15b. *H. mollis* var. *cordatus*

15a. **Helianthus mollis** Lam. var. **mollis** Fig. 72a.
Leaves rounded at the sessile base.

Common Name: Ashy sunflower; downy sunflower.
Habitat: Prairies, sand prairies, black oak barrens
Range: Maine to Wisconsin to Nebraska, south to Texas and Georgia.
Illinois Distribution: Common in most of the state.

This is a common variety found in prairies. It flowers from July to September.

15b. **Helianthus mollis** Lam. var. **cordatus** S. Wats. Gard. & Forest 2:136. 1889.
Fig. 72b.
Leaves cordate at the clasping base.
Except for the leaf differences, there seem to be no other significant differences
between the two varieties.
This variety flowers from July to September.

16. **Helianthus X cinereus** Torr. & Gray, Fl. N. Am. 2:324. 1842. Not illustrated.
Perennial herbs from rhizomes; stems erect, often unbranched, to 1.5 m tall,
hirsute or villous; leaves opposite, or the uppermost often alternate, lanceolate,
acute to acuminate at the apex, rounded or somewhat tapering to the sessile base,
to 18 cm long, to 4 cm wide, 3-nerved, usually serrulate, densely gray-pubescent,
glandular; heads radiate, solitary or up to 10, the pubescent peduncles up to 10 cm
long; involucre hemispheric to cylindric, 15–20 mm across; phyllaries 25–30, lanceolate, 8–12 mm long, 2.0–2.5 mm wide, usually pubescent; receptacle flat, paleate, the paleae 7–10 mm long; ray flowers 17–19, yellow, 1.8–2.2 cm long, neutral;
disc flowers 60–75, yellow, tubular, bisexual, fertile, the corolla 5-lobed, 5–6 mm
long, the disc 12–17 mm across; cypselae oblongoid, 3–4 mm long, glabrous except
for the pubescent apex; pappus of 2 aristate scales 2.0–2.5 mm long, occasionally
with tiny scales.

Common Name: Ashy sunflower; gray sunflower.
Habitat: Prairies.
Range: New York to Wisconsin, south to Texas and Georgia.
Illinois Distribution: Scattered in Illinois.

This is the hybrid between *H. mollis* and *H. occidentalis*. It resembles *H. mollis*, but
has smaller flowering heads with fewer rays and narrower leaves.
This hybrid flowers from July to October.

17. **Helianthus X brevifolius** E.E. Wats. Papers Mich. Acad. Sci. 9:448–449.
1929. Not illustrated.
Perennial herbs from rhizomes; stems erect, usually sparingly branched, to 2.5
m tall, sparsely gray-pubescent; leaves usually opposite below, alternate above,
lanceolate to lance-ovate, acute to acuminate at the apex, tapering or rounded at
the sessile or short-petiolate base, 20–25 cm long, 4–8 cm wide, usually serrate,

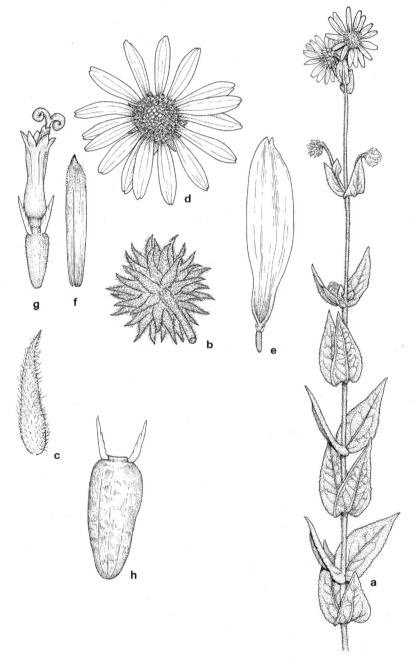

72a. *Helianthus mollis*
    var. *mollis* (Ashy
    sunflower).

a. Upper part of plant.
b. Cluster of phyllaries.
c. Phyllary.
d. Flowering head.

e. Ray flower.
f. Palea.
g. Disc flower.
h. Cypsela.

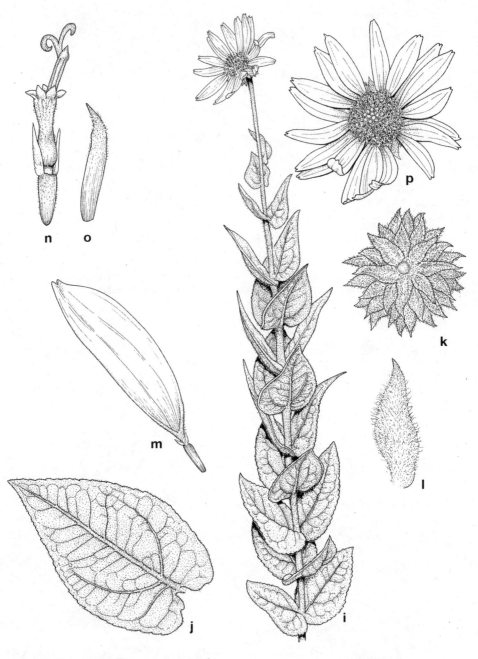

72b. *Helianthus mollis*
    var. *cordatus*
    (Ashy sunflower).
i. Upper part of plant.

j. Leaf.
k. Cluster of
   phyllaries.
l. Phyllary.

m. Ray flower.
n. Disc flower.
o. Palea.
p. Flowering head.

gray-pubescent on both surfaces, glandular; heads radiate, 5–15, on gray-pubescent peduncles to 12 cm long; involucre hemispheric, to 25 mm across; phyllaries up to 30, narrowly lanceolate, somewhat spreading, 12–15 mm long, 2.0–2.5 mm wide, ciliate; receptacle flat, paleate, the paleae 8–10 mm long; ray flowers 15–20, yellow, 2.5–3.0 cm along, neutral; disc flowers numerous, yellow, tubular, bisexual, fertile, the corolla 5-lobed, 5.5–6.5 mm long, the disc 18–25 mm across; cypselae oblongoid, 3–4 mm long, glabrous; pappus of 2 aristate scales 2.0–2.5 mm long.

Common Name: Short-leaved sawtooth sunflower.
Habitat: Mostly prairies, sometimes sandy.
Range: Maine to Minnesota, south to Texas and Georgia.
Illinois Distribution: Scattered in Illinois.

This is the hybrid between *H. mollis* and *H. grosseserratus*. It has the gray pubescence of *H. mollis*, but the narrower leaves of *H. grosseserratus*. It differs also from *H. grosseserratus* by its shorter leaves, hence its scientific and common names.
   Helianthus X brevifolius flowers from July to October.

   18. **Helianthus X laetiflorus** Pers. Syn. Pl. 2:476. 1807. Not illustrated.
   Perennial herbs with rhizomes; stems erect, usually branched, pubescent, to 2 m tall; leaves both basal and cauline, the cauline opposite except for the uppermost, lanceolate to narrowly ovate, acute to acuminate at the apex, tapering to the petiolate base, to 25 cm long, to 8 cm wide, serrate, hispid on both surfaces, glandular, the petioles to 8 cm long; heads radiate, few, on pubescent peduncles up to 12 cm long; involucre hemispheric, about 15 mm across; phyllaries 25–35, oblong-lanceolate, acute at the apex, unequal, to 12 mm long, 3.0–3.5 mm wide, usually puberulent; receptacle flat, paleate, the paleae about 10 mm long; ray flowers up to 20, yellow, to 3.5 cm long, neutral; disc flowers numerous, yellow or sometimes reddish-tipped, tubular, bisexual, sometimes fertile, the corolla 5-lobed, 7.0–7.5 mm long; cypselae usually not found, if present, oblongoid, 4–5 mm long, glabrous or nearly so; pappus of 2 aristate scales 2.5–3.5 mm long.

Common Name: Hybrid sunflower; cheerful sunflower.
Habitat: Black oak savannas, prairies, roadsides.
Range: Labrador to Minnesota, south to Texas and Georgia; also Montana and
   Oregon.
Illinois Distribution: Scattered in Illinois, but not common.

This plant is reputed to be the hybrid between *H. subrhomboideus* and *H. tuberosus*. It is distinguished by its pubescent stems, acute phyllaries, and distinct petioles. The cypselae are rarely formed.
   Helianthus X laetiflorus flowers during August and September.

   19. **Helianthus tuberosus** L. Sp. Pl. 2:905. 1753. Figs. 73a, 73b.
Helianthus tomentosus Michx. Fl. Bor. Am. 2:141. 1803.

Perennial herbs with rhizomes and tubers; stems erect, branched, to 3 m tall, hispid to hirsute, rarely glaucous; leaves basal and cauline, the cauline opposite or alternate, thick, firm, lance-ovate to ovate, acute to acuminate at the apex, tapering to the petiolate base, to 25 cm long, to 10 (–15) cm wide, 3-veined above the base, serrate or less often entire, scabrous on the upper surface, sparsely or densely pubescent on the lower surface, glandular, the petioles up to 5 cm long and winged; heads radiate, up to 15, on scabrous peduncles up to 15 cm long; involucre hemispheric, to 15 mm across; phyllaries up to 35, spreading to reflexed, lanceolate, to 15 mm long, 2–4 mm wide, more or less equal, pubescent; receptacle flat, paleate, the paleae 8–9 mm long; ray flowers up to 20, yellow, to 4 cm long, neutral; disc flowers numerous, yellow, tubular, bisexual, fertile, the corolla 5-lobed, 6–7 mm long, the disc about 15 mm across; cypselae oblongoid, 5–7 mm long, usually pubescent at the apex; pappus of 2 aristate scales 2–3 mm long, sometimes with a single tiny scale.

This handsome flowering species has underground tubers that may be eaten by humans. These account for the common name of Jerusalem artichoke. It occurs in a variety of native as well as disturbed habitats.

Two varieties occur in Illinois:

a. Upper and middle leaves alternate, moderately and inconspicuously pubescent on the lower surface with mostly appressed hairs. . . . . 19a. *H. tuberosus* var. *tuberosus*

a. All leaves opposite except for the uppermost two or three, pubescent on the lower surface with spreading hairs . . . . 19b. *H. tuberosus* var. *subcanescens*

19a. **Helianthus tuberosus** L. var. **tuberosus** Fig. 73a.

Upper and middle leaves alternate, moderately and inconspicuously pubescent on the lower surface with mostly appressed hairs.

Common Name: Jerusalem artichoke.

Habitat: Moist ground along roads, in fields, at the edges of woods.

Range: Nova Scotia to Saskatchewan and Washington, south to Utah, Texas, and Florida.

Illinois Distribution: Common throughout the state.

This typical variety flowers from July to October.

19b. **Helianthus tuberosus** L. var. **subcanescens** Gray, Syn. Pl. 1, part 2:280. 1884. Fig. 73b.

Helianthus subcanescens (Gray) E.E. Wats. Papers Mich. Acad. Sci. 9:430. 1929.

Helianthus mollissimus E.E. Wats. Papers Mich. Acad. Sci. 9:432. 1929.

All leaves opposite, except sometimes for the uppermost two or three, densely pubescent on the lower surface with spreading hairs.

This variety is found in similar habitats as var. *tuberosus*, and is scattered throughout the state.

It flowers from July to October.

73a. *Helianthus tuberosus*
  var. *tuberosus*
  (Jerusalem artichoke).

a. Upper part of plant.
b. Leaf.
c. Phyllary.
d. Flowering head.

e. Ray flower.
f. Palea.
g. Disc flower.
h. Cypsela.

73b. *Helianthus tuberosus*
var. *subcanescens*
(Jerusalem
artichoke).

a. Upper part of plant.
b. Node with opposite
   leaves.
c. Cluster of phyllaries.
d. Phyllary.

e. Flowering head.
f. Ray flower.
g. Palea.
h. Disc flower.
i. Cypsela.

20. **Helianthus hirsutus** Raf. Ann. Nat. 1:14. 1820. Figs. 74a, 74b, 74c.

Perennial herbs with rhizomes; stems erect, often unbranched, to 2 m tall, hirsute to hispid; leaves opposite, usually spreading, lanceolate to ovate, acute to acuminate at the apex, rounded at the sessile or short-petiolate base, to 15 cm long, to 9 cm wide, 3-nerved from the base, usually serrate, scabrous above, hispid to hirsute below, glandular; heads radiate, few, on scabrous peduncles up to 5 cm long; involucre hemispheric, to 25 mm across; phyllaries up to 25, spreading, lanceolate, to 12 mm long, 2.5–3.5 mm wide, ciliate; receptacle flat, paleate, the paleae up to 10 mm long; ray flowers up to 15 per head, yellow, to 2 cm long, neutral; disc flowers 40–50, yellow, tubular, bisexual, fertile, the corolla 5-lobed, 5.5–6.5 mm long, the disc 10–20 mm across; cypselae oblongoid, 4.0–4.5 mm long, usually pubescent at the apex; pappus of 2 aristate scales 2.5–3.0 mm long.

Three varieties occur in Illinois:
a. Some or all the leaves more than 2 cm wide, mostly spreading.
   b. Stems hirsute with long-spreading hairs . . . . 20a. *H. hirsutus* var. *hirsutus*
   b. Stems hispid with short stiff hairs . . . . 20b. *H. hirsutus* var. *trachyphyllus*
a. All leaves less than 2 cm wide, ascending. . . . . 20c. *H. hirsutus* var. *stenophyllus*

20a. **Helianthus hirsutus** Raf. var. **hirsutus** Fig. 74a.

Some or all the leaves more than 2 cm wide, up to 9 cm long, mostly spreading; stems hirsute with long-spreading hairs.

Common Name: Hairy sunflower.
Habitat: Dry woods, fields, savannas, prairies.
Range: New York to Minnesota, south to Texas and Florida.
Illinois Distribution: Scattered throughout the state.

This typical variety has densely hirsute stems and spreading leaves more than 2 cm wide.

Helianthus hirsutus var. *hirsutus* flowers from July to September.

20b. **Helianthus hirsutus** Raf. var. **trachyphyllus** Torr. & Gray, Fl. N. Am. 2:329. 1842. Fig. 74b.

Some or all the leaves more than 2 cm wide, up to 7 cm long, mostly spreading; stems hispid with short, stiff hairs.

This variety, scattered throughout the state, is the most common of the three varieties of *H. hirsutus* in Illinois.

Helianthus hirsutus var. *trachyphyllus* flowers from July to September.

20c. **Helianthus hirsutus** Raf. var. **stenophyllus** Torr. & Gray, Fl. N. Am. 2:329. 1842. Fig. 74c.

All the leaves less than 2 cm wide and ascending.

This narrow-leaved variety is found mostly in the southern half of Illinois.

Helianthus hirsutus var. *stenophyllus* flowers from July to September.

21. **Helianthus x doronicoides** Lam. Encycl. 3:84. 1789. Not illustrated.

Perennial herbs with rhizomes; stems erect, branched, stout, to 1.5 m tall,

74a. *Helianthus hirsutus* var. *hirsutus* (Hairy sunflower).

a. Upper part of plant.
b. Node with opposite leaves.
c. Phyllary.
d. Flowering head.
e. Ray flower.
f. Disc flower.
g. Cypsela.

rough-pubescent; leaves opposite, except sometimes for the uppermost, oblong to lance-ovate, broadest near the middle, acute to acuminate at the apex, tapering to the short-petiolate or nearly sessile base, to 18 cm long, to 6 cm wide, 3-nerved above the base, hispid and very scabrous on both surfaces, glandular, the petioles to 1 cm long, winged; heads radiate, up to 15, on hispid peduncles up to 12 cm long; involucre hemispheric, to 25 mm across; phyllaries 25–35, spreading, lanceolate, 10–15 mm long, 2–3 mm wide, pubescent; receptacle flat, paleate, the paleae 8–9 mm long; ray flowers 15–20, yellow, 2–3 cm long, neutral; disc flowers

74b. *Helianthus hirsutus*
    var. *trachyphyllus*
    (Hairy sunflower).
a. Upper part of plant.

b. Node with opposite
    leaves.
c. Cluster of phyllaries.
d. Phyllary.

e. Flowering head.
f. Ray flower.
g. Palea.
h. Disc flower.

numerous, yellow, tubular, bisexual, fertile, the corolla 5-lobed, 5.5–6.5 mm long, the disc 15–18 mm across; cypselae oblongoid, 3–4 mm long, glabrous; pappus of 2 aristate scales 2–3 mm long.

Common Name: Oblong-leaved sunflower.
Habitat: Dry woods.
Range: New Jersey to Minnesota, south to Texas and Arkansas.
Illinois Distribution: Known only from Sangamon County.

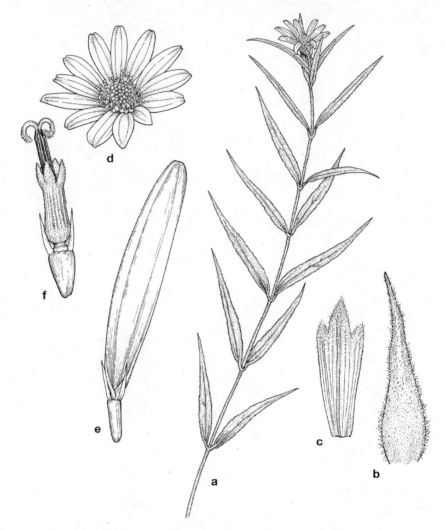

74c. *Helianthus hirsutus*
    var. *stenophyllus*
    (Hairy sunflower).

a. Upper part of plant.
b. Phyllary.
c. Palea.

d. Flowering head.
e. Ray flower.
f. Disc flower.

This plant is reputed to be the hybrid between *H. mollis* and *H. giganteus*. It resembles *H. hirsutus* var. *trachyphyllus* because of its hispid stems and mostly opposite leaves, but differs by its leaves broadest near the middle and its leaves that are 3-veined above the base.

    Helianthus X doronicoides flowers from July to October.

    22. Helianthus giganteus L. Sp. Pl. 2:905. 1753. Fig. 75.
    Perennial herbs with rhizomes; stems erect, stout, branched, to 3 m tall, hispid with spreading hairs, scabrous; leaves mostly alternate, or often the lowermost

opposite, lanceolate to narrowly ovate, acuminate at the apex, tapering to the short-petiolate or rarely sessile base, to 20 cm long, to 3.5 cm wide, 3-veined, serrulate or nearly entire, hirsute or hispid on both surfaces, glandular; heads radiate, up to 12 (–15), on scabrous peduncles up to 10 cm long; involucre hemispheric, 10–20 mm across; phyllaries up to 25, spreading to recurved, linear, to

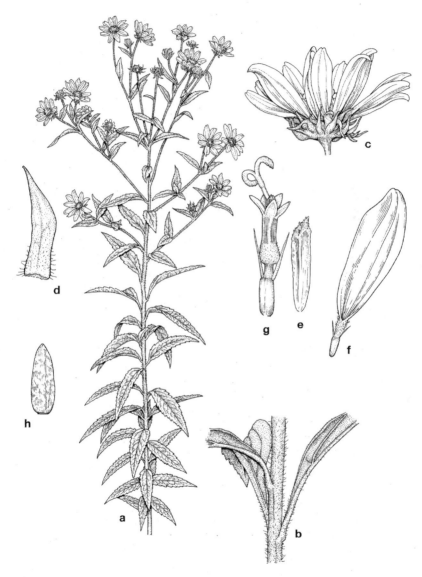

75. *Helianthus giganteus* (Giant sunflower).

a. Upper part of plant.
b. Node with leaf bases.
c. Flowering head.
d. Phyllary.

e. Palea.
f. Ray flower.
g. Disc flower.
h. Cypsela.

15 mm long, 1–2 mm wide, strigillose; receptacle flat, paleate, the paleae 7–9 mm long; ray flowers up to 15 (–20), yellow, 1.5–2.0 (–2.5) cm long, neutral; disc flowers numerous, yellow, tubular, bisexual, fertile, the corolla 5-lobed, 5–6 mm long, the disc 18–20 mm across; cypselae oblongoid, 3–4 mm long, pubescent at the apex; pappus of 2 aristate scales 2.5–3.0 mm long.

Common Name: Giant sunflower.
Habitat: Sandy prairies, fens.
Range: Newfoundland to Minnesota, south to Mississippi and Georgia; also Kansas.
Illinois Distribution: Known from Cook, Kane, Tazewell, and Winnebago counties.

This robust species is similar in appearance and stature to *H. grosseserratus*, but differs by its hispid and scabrous stems.

Helianthus giganteus flowers from August to October.

23. **Helianthus X intermedius** R. L. Long, Rhodora 56:201. 1954. Not illustrated.
Perennial herbs with rhizomes; stems stout, erect, usually unbranched, to 3 m tall, pubescent with short, white hairs up to 1 mm long, scabrous at least at the top; leaves alternate, lanceolate, acuminate at the apex, tapering to the short-petiolate base, to 20 cm long, to 3.5 cm wide, sometimes conduplicate, with short serrations or sometimes nearly entire, with numerous short hairs on the lower surface, the petioles 5–15 mm long; heads radiate, up to 12 in spikes, on scabrous peduncles to 8 cm long; involucre hemispheric, 12–25 mm across; phyllaries various, often spreading, lanceolate, 12–15 mm long, 2–3 mm wide, gray-pubescent, ciliate; receptacle flat, paleate, the paleae 7–9 mm long; ray flowers up to 25, yellow, 2.0–3.5 mm long, neutral; disc flowers numerous, yellow, tubular, bisexual, the corolla 5-lobed, to 6 mm long, the disc 15–25 mm across; cypselae not seen.

Common Name: Hybrid sunflower.
Habitat: Disturbed wetland.
Range: Ohio to Minnesota, south to Texas.
Illinois Distribution: Known only from DuPage County.

This is the reputed hybrid between *Helianthus giganteus* L. and *Helianthus maximilianii* Schrad.

It has spreading hairs like those in *H. giganteus*, and some of the leaves are conduplicate and the phyllaries are ciliate like those in *H. maximilianii*.

This hybrid flowers in August and September.

24. **Helianthus maximilianii** Schrad. Index Sem. Hort. Goett. 1835. Figs. 69b (k–m), 76.
Perennial herbs with rhizomes; stems often several, stout, erect, mostly un-branched, to 4 m tall, pubescent with appressed hairs, usually scabrous; leaves alternate, lanceolate, acuminate at the apex, tapering to the short-petiolate or nearly sessile base, to 30 cm long, to 5 cm wide, conduplicate, not 3-veined, entire

76. *Helianthus maximilianii* (Maximilian's sunflower).

a. Upper part of plant.
b. Stem with cauline leaves.
c. Cluster of phyllaries.
d. Phyllary.

e. Flowering head.
f. Ray flower.
g. Palea.
h. Disc flower.
i. Cypsela.

or less commonly serrulate, scabrous, sometimes hispid, glandular; heads radiate, up to 15 in racemes or spikes, on scabrous peduncles up to 10 cm long; involucre hemispheric, 15–30 mm across; phyllaries 30–40, spreading, sometimes squarrose, lanceolate, 15–20 mm long, 2–3 mm wide, gray-pubescent, ciliate; receptacle flat, paleate, the paleae 8–10 mm long; ray flowers up to 30, yellow, 2.5–4.0 cm long, neutral; disc flowers numerous, yellow, tubular, bisexual, fertile, the corolla 5-lobed, 5–7 mm long, the disc 20–30 mm across; cypselae narrowly oblongoid, 3–4 mm long, glabrous or nearly so; pappus of 2 aristate scales 3–4 mm long.

Common Name: Maximilian's sunflower.
Habitat: Naturalized in prairies and along roads.
Range: Native to the western United States; usually adventive east of the Mississippi River.
Illinois Distribution: Occasional and scattered in the state.

This tall, many-stemmed species is handsome because of its several flowering heads in racemes or spikes. The leaves are all alternate, have only one main vein, and are usually conduplicate.

This species, not native to Illinois, flowers in July and August.

25. **Helianthus X luxurians** E.E. Wats. Papers Mich. Acad. Sci. 9:464–465. 1929. Not illustrated.

Perennial herbs with rhizomes; stems stout, erect, usually branched, to 3.5 m tall, scabrous but not pubescent; leaves alternate, except usually for the lowest few pairs, lanceolate to narrowly ovate, acuminate at the apex, tapering to the short-petiolate base, to 25 cm long, to 6 cm wide, 3-veined, serrate or sometimes nearly entire, scabrous above and below, glandular; heads radiate, 12–18, on usually scabrous peduncles up to 12 cm long; involucre hemispheric, 15–25 mm across; phyllaries 25–30, recurved, narrowly lanceolate, glabrous or puberulent, ciliate, 10–15 mm long, 1.0–2.2 mm wide; receptacle flat, paleate, the paleae 7–9 mm long; ray flowers 10–20, yellow, 2.0–3.5 cm long, neutral; disc flowers numerous, yellow, tubular, bisexual, fertile, the corolla 5-lobed, 5–6 mm long, the disc 15–25 mm across; cypselae oblongoid, 3–4 mm long, glabrous or sometimes pubescent at the apex; pappus of 2 aristate scales 2–3 mm long.

Common Name: Luxuriant sunflower.
Habitat: Sand prairies, fens.
Range: Quebec to Minnesota, south to Texas and Georgia.
Illinois Distribution: Occasional in the state.

This is reputed to be the hybrid between *H. grosseserratus* and *H. giganteus*. It resembles both of them in stature and most other characteristics. It has scabrous stems, whereas *H. grosseserratus* has glabrous stems and *H. giganteus* has hairy stems.

Helianthus X luxurians flowers during August and September.

Subtribe Galinsoginae Benth.

Annuals, perennials, or shrubs; stems usually erect, branched or unbranched; leaves opposite (in Illinois) or alternate, simple, petiolate or sessile, sometimes gland-dotted; heads radiate (in Illinois) or discoid, borne singly, in corymbs, or in cymes; involucre campanulate (in Illinois) or hemispheric or obpyramidal; phyllaries up to 30 in 2–5 series, usually persistent, mostly unequal; receptacle conic (in Illinois), flat, or convex, paleate; ray flowers up to 8 (in Illinois) or up to 20 or absent, white or yellow or pink (in Illinois), pistillate, fertile; disc flowers 50-numerous, tubular, bisexual, fertile, pink, yellow, or white (in Illinois), the corolla 5-lobed; cypselae flat, 3- to 5-angled; pappus of laciniate or fimbriate scales, or absent.

Fifteen genera are in this subtribe, with about a hundred species, all in warmer parts of the New World. Only the following genus occurs in Illinois.

53. **Galinsoga** Ruiz & Pavon—Peruvian Daisy; Quickweed.

Annual herbs with fibrous roots; stems ascending to erect; leaves opposite, simple; flowering heads radiate (in Illinois) or discoid, borne in cymes; involucre campanulate or hemispheric; phyllaries in 2 series, more or less equal, obtuse, fringed (in Illinois); receptacle conic, paleate; ray flowers up to 15 (in Illinois), rarely absent, white to pinkish, pistillate, fertile; disc flowers up to 150, tubular, yellow, perfect, fertile, the corolla 5-lobed; cypselae flat, obconic to obpyramidal, glabrous or pubescent; pappus of up to 20 short bristles, or absent.

Galinsoga consists of about thirty species, mostly native to tropical America. Two species occur in Illinois, both of them introduced.

1. Pappus of disc flowers conspicuously fringed, not tapering to awn tips, persistent; pappus of ray flowers absent or minute; outer phyllaries 2–4. . . . . 1. *G. parviflora*
1. Pappus of disc flowers slightly fringed, tapering to an awn tip, caducous; pappus of ray flowers well developed; outer phyllaries 1–2 . . . . 2. *G. quadriradiata*

1. **Galinsoga parviflora** Cav. Icon. 3:41. 1794. Fig. 77.

Annual herbs with fibrous roots; stems branched from the base, ascending to erect, to 75 cm tall, glabrous or with appressed pubescence; leaves opposite, simple, lance-ovate to ovate, acute to acuminate at the apex, rounded to or tapering at the base, to 7.5 cm long, to 3.5 cm wide, shallowly dentate to nearly entire, 3-nerved from the base, appressed-pubescent to nearly glabrous, the lowermost petiolate, the upper smaller and sessile; flowering heads radiate, 4–6 mm across, on rather stiff peduncles up to 4 cm long; involucre campanulate, 2.5–5.0 mm across; phyllaries in 2 series, the outer 2–4 in number, the inner more numerous, persistent, glabrous or nearly so, obtuse to acute, slightly fringed; receptacle conic, paleate; ray flowers radiate, up to 15 per head, up to 2 mm long, white or pinkish, pistillate, fertile; disc flowers up to 50, tubular, whitish, bisexual, fertile, the corolla 5-lobed; cypselae obpyramidal, pubescent or rarely glabrous, 1.5–2.5 mm long; pappus of ray flowers of 5–10 minute scales, or absent; pappus of disc flowers of up to 20 linear scales up to 2 mm long, conspicuously fringed, not tapering to an awned tip, persistent, or absent.

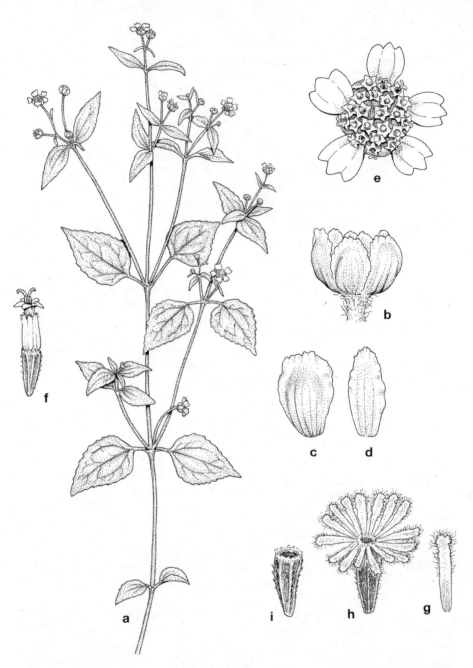

77. *Galinsoga parviflora*
   (Lesser Peruvian daisy).
a. Upper part of plant.
b. Cluster of outer
   phyllaries.

c. Outer phyllary.
d. Inner phyllary.
e. Flowering head.
f. Disc flower.

g. Palea.
h. Cypsela from disc
   flower.
i. Cypsela from ray flower.

Common Name: Lesser Peruvian daisy, smooth Peruvian daisy.
Habitat: Disturbed soil.
Range: Native to tropical America; adventive throughout most of the United States, except for the southeastern states.
Illinois Distribution: Occasional in the northern half of the state; also Jackson County.

This is a small, innocuous weed that lives in disturbed soil, primarily in towns and cities.

It differs from *G. quadriradiata* by the usual absence of pappus on the cypselae of the ray flowers, the conspicuously fringed pappus of the disc flower, and the shallowly dentate to entire leaves.

This species flowers from June to October.

2. **Galinsoga quadriradiata** Ruiz & Pavon, Syst. Veg. Fl. Peruv. & Chile 1:198. 1798. Fig. 78.
Adventina ciliata Raf. New Fl. 1:67. 1836.
Galinsoga ciliata (Raf.) S.F. Blake, Rhodora 24:35. 1922.

Annual herbs with fibrous roots; stems branched from the base, erect, hispid, often glandular, to 50 cm tall; leaves opposite, simple, ovate, acute to acuminate at the apex, rounded at or tapering to the base, to 6 cm long, to 3.5 cm wide, dentate, 3-nerved from the base, hispid; flowering heads radiate, 6–8 mm across, on stiff peduncles up to 2 cm long; involucre campanulate to hemispheric, 3–6 mm across; phyllaries in 2 series, the outer 1–2, ovate, ciliate, caducous; receptacle conic, paleate; ray flowers up to 8 per head, to 2.5 mm long, white or less commonly pink, pistillate, fertile; disc flowers up to 35, white, tubular, bisexual, fertile, the corolla 5-lobed; cypselae obpyramidal, hispidulous, those of the ray flowers 1.5–2.0 mm long, those of the disc flowers up to 1.8 mm long; pappus of cypselae of ray flowers well-developed, of 6–15 fimbriate scales up to 1 mm long, the pappus of cypselae of the disc flowers up to 20, fimbriate, aristate, to 1.7 mm long.

Common Name: Peruvian daisy; quickweed.
Habitat: Disturbed soil.
Range: Native to tropical America; adventive in much of the United States, except for the southeastern states.
Illinois Distribution: Occasional throughout Illinois.

This is the more common of the two species of *Galinsoga* in Illinois. It differs from *G. parvifolia* by its usually larger stature and the aristate awns of the pappus of the cypselae of the disc flowers. For many years it was known as *G. ciliata*.

This species occurs in disturbed soil. In the northern half of the state, it is found primarily in cities and towns, but in southern Illinois, it grows at the edge of disturbed wetlands.

Galinsoga quadriradiata flowers from June to November.

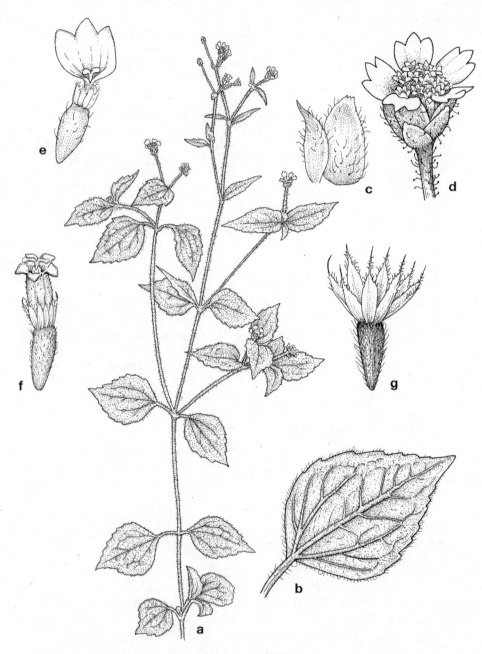

78. *Galinsoga*
   quadriradiata
   (Peruvian daisy).

a. Upper part of plant.
b. Leaf.
c. Inner and outer
   phyllaries.

d. Flowering head.
e. Ray flower.
f. Disc flower.
g. Cypsela.

Subtribe Coreopsidinae Less.

Annuals, perennials, or shrubs; leaves basal or cauline, the cauline mostly opposite, simple or compound; heads radiate or discoid, borne in corymbs, panicles, or
cymes, subtended by bractlets as well as phyllaries; involucre campanulate, cylindric, or hemispheric; phyllaries up to 40, in 2 series, persistent, usually scarious;
receptacle flat or convex, paleate; ray flowers up to 20, pistillate, fertile, or neutral,
usually yellow, sometimes absent; disc flowers numerous, tubular, bisexual, fertile,
the corolla 4- or 5-lobed; cypselae more or less flattened, sometimes angled, sometimes narrowly winged, glabrous or pubescent; pappus usually of 2–8 awns, rarely
reduced to a crown of scales.

This subtribe consists of 20 genera and 375 species found in subtropical and
warm temperate regions in both the Old and New Worlds. Seven genera and seventy
species occur in the United States. There are four genera of this subtribe in Illinois.

1. Cypselae angular; phyllaries free from each other.
    2. Leaves usually 3-pinnate, less commonly 2-pinnate . . . . 56. *Cosmos*
    2. Leaves simple or 1-pinnate (2-pinnate in *Bidens bipinnata*), or palmately divided.
        3. Cypselae with 2–4 stiff awns . . . . 57. *Bidens*
        3. Cypselae with 2 or more weak awns or scales . . . . .54. *Coreopsis*
1. Cypselae terete; inner phyllaries connate below . . . . 55. *Thelesperma*

54. **Coreopsis** L.—Tickseed Sunflower; Coreopsis

Annual or perennial herbs (in Illinois); stems ascending to erect, usually
branched; leaves basal and/or cauline, the cauline mostly opposite or the uppermost sometimes alternate, simple or pinnately divided or palmately divided; heads
radiate, borne in corymbs, occasionally borne singly, subtended by a few bractlets;
involucre hemispheric or cylindric; phyllaries in 2 series, the outer narrower than
the inner, usually scarious along the margins, usually with small bractlets, called
calyculi, at their base; receptacle flat to convex, paleate; ray flowers usually 8 per
head, usually yellow, pistillate, fertile, sometimes neutral; disc flowers yellow,
tubular, bisexual, fertile, the corolla 4- or 5-lobed; cypselae flat, orbicular, oblong,
or linear, usually narrowly winged; pappus of 2 very short awns or teeth.

About thirty-six species comprise this genus, mostly in temperate North
America. Nine species occur in Illinois. Several species are cultivated because of
their attractive flowering heads. Sometimes the rays are "doubled" in horticultural
varieties.

North American species of *Coreopsis* may be divided into eight sections, three of
which occur in Illinois. Section *Gyrophyllum* in Illinois consists of *C. palmata* and
C. tripteris. This section is characterized by having the lobes of the disc corollas
5 in number, rays that are broadest at or near the middle, with the paleae usually branched at the tip. Section *Coreopsis* has six species in Illinois—*C. basalis, C.
lanceolata, C. crassifolia, C. grandiflora, C. verticillata,* and *C. pubescens*. The species in
this section have disc corollas with 5 lobes, rays that are broadest above the middle, and paleae that are filiform at the tip. Section *Caliopsis* is represented in Illinois
only by *C. tinctoria*. The lobes of the disc corolla in this section are 4 in number.

The second edition of Britton and Brown's Illustrated Flora of the Northeastern United States attributes *C. auriculata* L. to Illinois, but there are apparently no confirming specimens.

1. Leaves undivided, or with 1–2 short lateral lobes.
  2. Leaves mostly basal, linear to oblanceolate to obovate.
    3. Stems glabrous or nearly so; leaves glabrous; nodes on aerial stems 1–5, the internodes up to 5 cm long.................................... 4. *C. lanceolata*
    3. Stems spreading-villous; leaves pubescent; nodes on aerial stems more than 5, the internodes 6 cm long or longer........................... 5. *C. crassifolia*
  2. Leaves developed to middle of stem or higher, ovate to elliptic to lanceolate ....
    .... 7. *C. pubescens*
1. Leaves 3- to 5-lobed or–divided.
  4. Some or all the leaves or leaflets 3 mm wide or wider.
    5. Leaves sessile, deeply 3- or 5-lobed below the middle.............. 1. *C. palmata*
    5. Leaves usually petiolate, divided into 3–5 segments.
      6. Ligules of ray flowers reddish brown at base or throughout; disc flowers reddish brown.
        7. Cypselae wingless; leaf segments linear to linear-lanceolate... 8. *C. tinctoria*
        7. Cypselae narrowly winged; leaf segments lanceolate to orbicular ....
        .... 3. *C. basalis*
      6. Ligules of ray flowers entirely yellow; disc flowers yellow or reddish brown.
        8. Leaf segments elliptic-lanceolate; cypselae 5–7 mm long; disc flowers yellow or reddish brown.. 2. *C. tripteris*
        8. Leaf segments linear-filiform to linear-lanceolate; cypselae 2–4 mm long; disc flowers yellow................................... 6. *C. grandiflora*
  4. All leaflets up to 2 mm wide............................... 9. *C. verticillata*

    1. **Coreopsis palmata** Nutt. Gen. N. Am. Pl. 2:180. 1818. Fig. 79.

Perennial herbs with rhizomes; stems erect, usually unbranched, to 75 cm tall, glabrous or pubescent only at the nodes; leaves opposite, or the uppermost sometimes alternate, deeply palmately 3- or 5-lobed, the lobes linear, more or less obtuse at the apex, entire, glabrous or nearly so, to 4.5 cm long, sessile; flowering heads radiate, on peduncles to 4 cm long, subtended by up to 12 bractlets; involucre hemispheric; phyllaries in 2 series, the outer narrower and little shorter than the inner, to 9 mm long; receptacle flat to convex, paleate; ray flowers usually 8 in number, bright yellow, oblong, 3-notched at the apex, up to 2.5 cm long, pistillate, usually fertile; disc flowers up to 80 per head, yellow, tubular, bisexual, fertile, the corolla 5-lobed, 5–6 mm long; cypselae flat, oblong, narrowly winged, glabrous or nearly so, to 6 mm long; pappus of 2 short teeth.

Common Name: Prairie coreopsis.
Habitat: Prairies, open woods, black oak savannas.
Range: Wisconsin to South Dakota, south to Oklahoma, Louisiana, and Indiana.
Illinois Distribution: Common in the northern three-fourths of Illinois, occasional elsewhere.

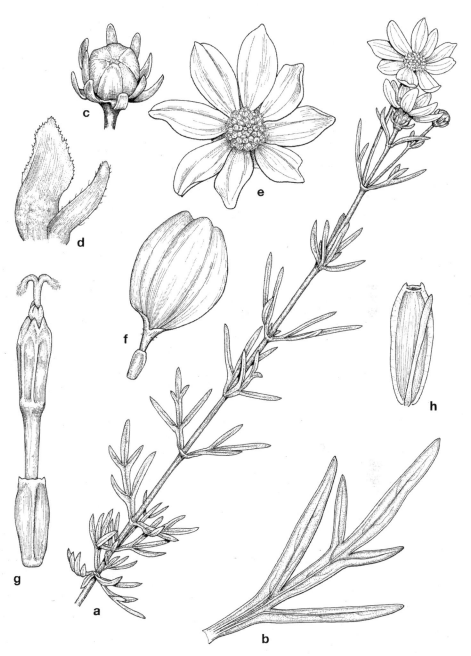

79. *Coreopsis palmata*
(Prairie coreopsis).
a. Upper part of plant.
b. Leaf.

c. Unopened flowering
   head.
d. Outer and inner
   phyllaries.

e. Flowering head.
f. Ray flower.
g. Disc flower.
h. Cypsela.

This species is recognized by its sessile, palmately 3- (−5-) lobed opposite leaves.

Although fairly common in prairies, this species is also characteristic of black oak savannas in the northern half of Illinois.

Coreopsis palmata flowers from June to August.

2. Coreopsis tripteris L. Sp. Pl.2:908. 1753. Figs. 80a, 80b, 80c.

Perennial herbs with rhizomes; stems ascending to erect, usually unbranched, to 3 m tall, hirsute or villous, at least on the lowest part of the stem, or glabrous; leaves usually opposite, except sometimes for the uppermost, simple and lanceolate or more commonly palmately 3- or 5-parted, to 8 cm long, the divisions linear-lanceolate to linear, glabrous or pubescent, entire, acute at the apex; flowering heads radiate, numerous, on peduncles up to 6 cm long, with 5–6 bractlets at the base up to 5 mm long; involucre hemispheric; phyllaries in 2 series, the outer linear, more or less obtuse, shorter than the inner ones, the inner narrowly ovate, acute at the apex, glabrous or pubescent; receptacle flat, paleate; ray flowers usually 8 per head, yellow, obtuse and not notched at the apex, to 2 cm long, pistillate, fertile; disc flowers up to 80, tubular, reddish brown or yellow, bisexual, fertile, the corolla 5-lobed, 5–6 mm long; cypselae flat, oblong to obovate, 5–7 mm long, glabrous or nearly so, papillose; pappus absent.

Three varieties occur in Illinois:
a. All leaves compound.
  b. Lower surface of leaves and outer phyllaries glabrous or nearly so . . . .
  . . . . .2a. *C. tripteris* var. *tripteris*
  b. Lower surface of leaves and outer phyllaries densely pubescent . . . .
  . . . . 2b. *C. tripteris* var. *deamii*
a. Most of the leaves simple. . . . . 2c. *C. tripteris* var. *intercedens*

2a. Coreopsis tripteris L. var. tripteris Fig. 80a.

All leaves compound; lower surface of leaves and outer phyllaries glabrous or nearly so.

Common Name: Tall tickseed.
Habitat: Dry woods, sandy woods, prairies.
Range: Quebec to Wisconsin, south to Texas and Florida.
Illinois Distribution: Common throughout the state.

This is the more common of the three varieties in Illinois. It flowers from July to October.

2b. Coreopsis tripteris L. var. deamii Standl. Rhodora 32:33–34. 1930. Fig. 80b.

All leaves compound; lower surface of leaves and phyllaries densely pubescent.

Common Name: Deam's tall tickseed.
Habitat: Dry woods, prairies.
Range: Pennsylvania to Missouri, south to Arkansas and Alabama.

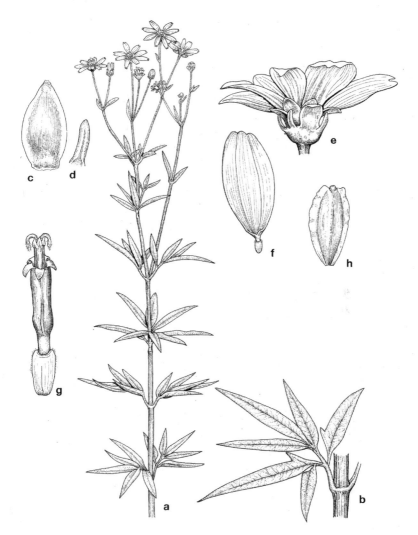

80a. *Coreopsis tripteris* var. *tripteris* (Tall tickseed).

a. Upper part of plant.
b. Node with leaf.
c. Outer bract.
d. Inner bract.

e. Flowering head.
f. Ray flower.
g. Disc flower.
h. Cypsela.

Illinois Distribution: Scattered throughout the state; apparently not quite as common as the typical variety.

The compound leaves and pubescent leaves and phyllaries are very distinctive for this variety.

The type for var. *deamii* was collected by H. N. Patterson in Henderson County in 1871.

Variety *deamii* flowers from July to October.

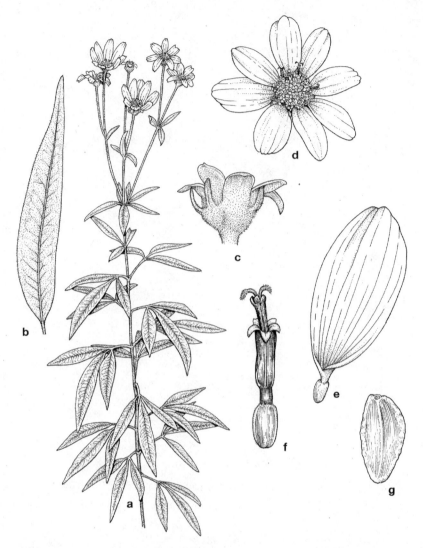

80b. *Coreopsis tripteris*
     var. *deamii*
     (Deam's tall
     tickseed).

a. Upper part of plant.
b. Leaf.
c. Cluster of phyllaries.
d. Flowering head.

e. Ray flower.
f. Disc flower.
g. Cypsela.

2c. **Coreopsis tripteris** L. var. **intercedens** Standl. Rhodora 32:3. 1930. Fig. 80c.
Most or all the leaves simple.

*Common Name*s: Tall tickseed; tall coreopsis.
Habitat: Dry woods, prairies.
Range: Quebec to Wisconsin, south to Texas and Florida.
Illinois Distribution: Scattered throughout the state.

The most striking difference about this variety is the lack or near lack of divided leaves. The type for var. *intercedens* was collected by F. C. Gates at Edgewater, Chicago, Cook County.

This variety flowers from July to October.

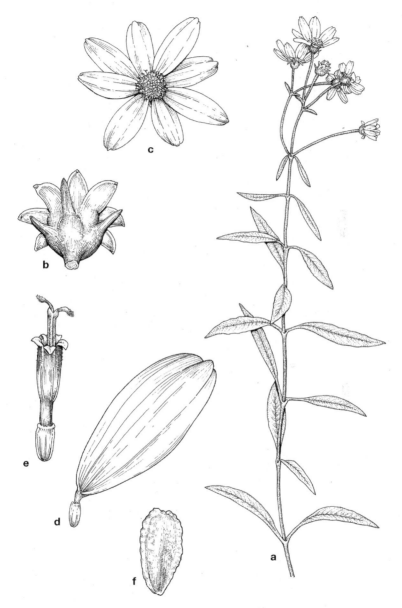

80c. *Coreopsis tripteris* var. *intercedens* (Tall tickseed).

a. Upper part of plant.
b. Cluster of phyllaries.
c. Flowering head.

d. Ray flower.
e. Disc flower.
f. Cypsela.

3. **Coreopsis basalis** (A. Dietr.) S.F. Blake, Proc. Am. Acad. Arts 51:525. 1916. Fig. 81.

Annual herbs with taproots; stems erect, to 50 cm tall, glabrous, usually unbranched; leaves basal and cauline, opposite, simple, usually 3- to 9-lobed, to 5 cm long, to 1.5 cm wide, the lobes elliptic to lanceolate to orbicular, glabrous, on petioles up to 8 cm long; heads radiate, on peduncles up to 15 cm long, with several linear bractlets up to 12 mm long; involucre hemispheric; phyllaries in 2 series, lance-ovate, the outer about half as long to nearly as long as the inner, glabrous or puberulent; receptacle flat, paleate, the paleae with a filiform tip; ray flowers usually 8, yellow, with a purplish or red-brown spot, to 2 cm long, pistillate, fertile; disc flowers up to 75, tubular, red-brown to purple, bisexual, fertile, the corolla 5-lobed, 3–4 mm long; cypselae more or less obovate, 1.2–1.8 mm long, narrowly winged, glabrous or nearly so; pappus absent.

Common Name: Ornamental tickseed; ornamental coreopsis.
Habitat: Disturbed soil in a pasture.
Range: Native to the southwestern United States; rarely escaped from cultivation.
Illinois Distribution: Known only from Lake County.

The distinctive characteristics of this species are its pinnately lobed leaves, its rays with a reddish or purplish base, and its narrowly winged cypselae. It is similar to *C. tripteris* but has broader leaf segments and broader winged cypselae.

This ornamental flowers from June to September.

4. **Coreopsis lanceolata** L. Sp. Pl. 2:1753. Fig. 82.
Coreopsis lanceolata L. var. *angustifolia* Torr. & Gray, Fl. N. Am. 2:344. 1942.

Perennial herbs with a short woody caudex; stems erect, to 60 cm tall, usually unbranched, glabrous or short-hispid near the base, with up to 3 nodes, the internodes up to 5 cm long; leaves basal or mostly near the base of the stem, opposite, simple, sometimes slightly lobed, the lower ones narrowly oblong to spatulate, obtuse at the apex, short-petiolate, the upper lanceolate to oblong, obtuse to subacute at the apex, sessile or subsessile, glabrous or nearly so, to 12 cm long, to 1.5 cm wide; head radiate, 1.8–3.5 cm across, solitary on a naked glabrous peduncle up to 25 cm long, subtended by linear to lanceolate bractlets up to 10 mm long; involucre hemispheric; phyllaries in 2 series, the outer narrower but nearly as long as the inner; receptacle flat, paleate, the paleae 4–6 mm long; ray flowers usually 8, yellow, 3- to 5-notched at the tip, up to 3 cm long, pistillate, fertile; disc flowers up to 80, tubular, yellow, the corolla 5-lobed, the disc 1.5–2.0 cm across; cypselae more or less orbicular, narrowly winged, 2.5–3.5 mm long; pappus of two short triangular teeth.

Common Name: Sand coreopsis.
Habitat: Sandy soil, rocky soil, dry prairies, black oak savannas.

81. *Coreopsis basalis*
(Ornamental tickseed).

a. Upper part of plant.
b. Cluster of phyllaries.

c. Disc flower.
d. Section of cypsela.

Range: Vermont to Wisconsin, south to Nebraska, New Mexico, Texas, and Florida.
Illinois Distribution: Scattered throughout the state.

This species is similar to *C. crassifolia*, but the aerial stems have about three inter-nodes that are usually 1–5 cm long, the stems are glabrous or nearly so, and the lower leaves are narrower and more obtuse.

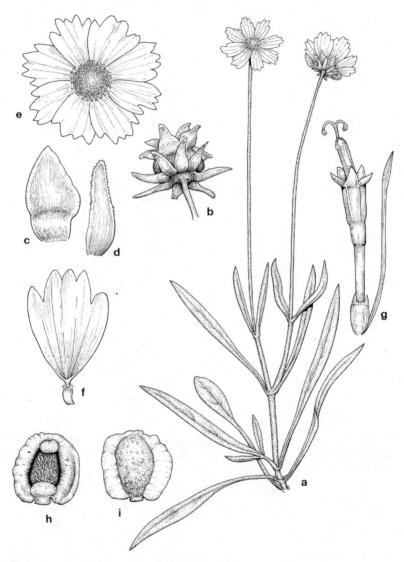

82. *Coreopsis lanceolata*
   (Sand coreopsis).
a. Upper part of plant.
b. Cluster of phyllaries.

c. Outer phyllary.
d. Inner phyllary.
e. Flowering head.
f. Ray flower.

g. Disc flower with
   subtending bract.
h, i. Cypselae.

This is a species of dry habitats and is often grown as an ornamental. Flowering time is April to August.

5. **Coreopsis crassifolia** Ait. Hort. Kew. 3:253. 1789. Fig. 83.
Coreopsis lanceolata L. var. *villosa* Michx. Fl. Bor. Am. 2:137. 1803.

Perennial herbs with a short woody caudex; stems erect, to 60 cm tall, usually unbranched, spreading-villous, with 5 or more nodes, the internodes at least 6 cm long; leaves basal or mostly on the lower part of the stem, opposite, simple, usually unlobed, the basal leaves oblong, obtuse to subacute at the apex, short-petiolate, the upper leaves oblong to obovate, subacute at the apex, short-petiolate to sessile, pubescent on both surfaces, to 8 cm long, to 1.2 cm wide; head radiate, solitary on a naked peduncle up to 25 cm long, 1.5–3.0 cm across, subtended by linear-lanceolate bractlets up to 10 mm long; involucre hemispheric; phyllaries in 2 series, the outer narrower and nearly as long as the inner; receptacle flat, paleate, the paleae 4–6 mm long; ray flowers usually 8, yellow, usually 5-notched at the tip, up to 2.7 cm long, pistillate, fertile; disc flowers up to 80, tubular, yellow, bisexual, fertile, the corolla 5-lobed, the disc 1.2–1.8 cm across; cypselae orbicular, narrowly winged, 2.5–3.0 mm long; pappus of 2 short terminal teeth.

Common Name: Hairy sand coreopsis.
Habitat: Sandy soil, dry prairies.
Range: Ontario to Missouri, south to Louisiana and Florida.
Illinois Distribution: Scattered in Illinois.

This species is often considered a variety of *C. lanceolata*. It differs by its more numerous nodes and longer internodes and its spreading-villous stems and leaves. It is often cultivated as an ornamental.

Coreopsis crassifolia flowers from April to August.

6. **Coreopsis grandiflora** Hogg ex Sweet, Brit. Fl. Gard. 2:plate 175. 1826. Figs. 84a, 84b.
Perennial herbs with a short woody caudex; stems erect, branched, to 1 m tall, glabrous; leaves cauline, opposite, the lower ones sometimes entire and petiolate, the middle and upper ones pinnately 1- to 2-divided, the 5–9 segments linear to linear-lanceolate to linear-filiform, to 45 cm long, the segments to 5 mm wide, glabrous; heads radiate, 1 to several in corymbs, 4–7 cm across, on naked glabrous peduncles up to 15 cm long, subtended by linear to lanceolate bractlets up to 10 mm long; involucre hemispheric; phyllaries in 2 series, the outer linear-subulate, the inner narrowly ovate, ciliate, to 10 mm long; receptacle flat, paleate, the paleae 6–7 mm long; ray flowers usually 8, yellow, 3- or 5-notched at the apex, to 25 mm long, pistillate, fertile; disc flowers up to 80, tubular, yellow, bisexual, fertile, the corolla 5-lobed, 3.5–4.5 mm long; cypselae broadly elliptic, broadly winged, 2–4 mm long; pappus of 2 short scales or absent.

Two varieties occur in Illinois:

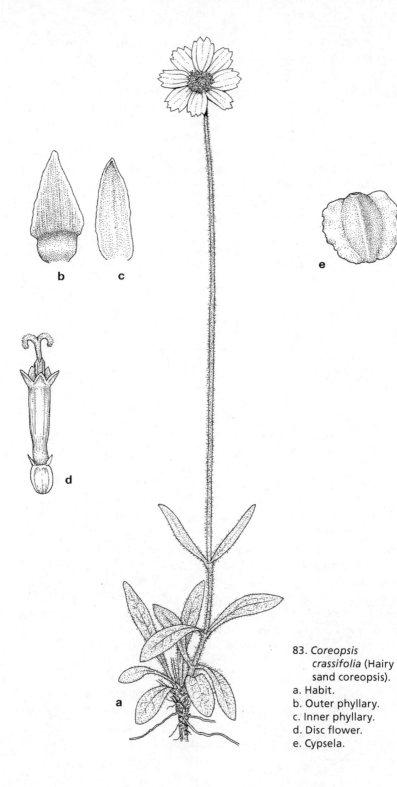

83. *Coreopsis crassifolia* (Hairy sand coreopsis).
a. Habit.
b. Outer phyllary.
c. Inner phyllary.
d. Disc flower.
e. Cypsela.

a. Leaves linear-lanceolate . . . . 6a. *C. grandiflora* var. *grandiflora*

a. Leaves linear-filiform . . . . 6b. *C. grandiflora* var. *harveyana*

6a. Coreopsis grandiflora Hogg ex Sweet var. **grandiflora** Fig. 84a.
Leaves linear-lanceolate.

Common Name: Large-flowered coreopsis.
Habitat: Disturbed soil.

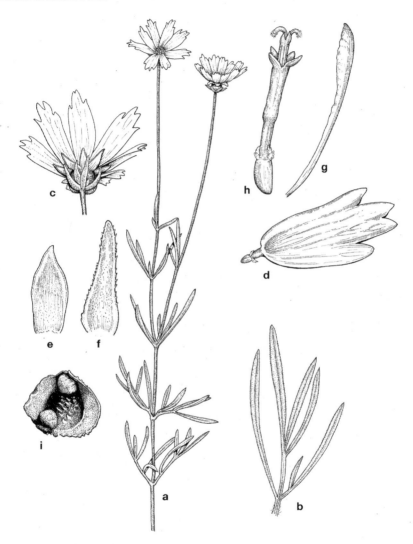

84a. *Coreopsis grandiflora*   a. Upper part of plant.   e, f. Phyllaries.
     var. *grandiflora*   b. Nodes with leaves.   g. Palea.
     (Large-flowered   c. Flowering head.   h. Disc flower.
     coreopsis).   d. Ray flower.   i. Cypsela.

Range: Native to the southeastern United States; occasionally escaped from cultivation.
Illinois Distribution: Scattered in Illinois.

This variety is a little less common in Illinois than var. *harveyana*. It is sometimes grown as an ornamental.

Variety *grandiflora* flowers from June to August.

6b. **Coreopsis grandiflora** Hogg ex Sweet var. **harveyana** (Gray) Sherff, Bot.Gaz. 94:593. 1933. Fig. 84b.
Coreopsis harveyana Gray, Syn. Fl. N. Am. 1:292. 1884.

Leaves linear-filiform.

Common Name: Narrow-leaved large-flowered coreopsis.
Habitat: Disturbed soil.
Range: Illinois and Missouri, south to Oklahoma and Arkansas.
Illinois Distribution: Scattered in Illinois.

The linear-filiform leaves distinguish this variety from typical var. *grandiflora*. It is sometimes grown as an ornamental.

This variety flowers from June to August.

7. **Coreopsis pubescens** Ell. Bot. S.C. & Ga. 2:441. 1824. Fig. 85.
Perennial herbs with a short woody caudex; stems erect, mostly unbranched, to 1.2 m tall, pubescent; lower leaves opposite, oval to obovate, simple or with 3 lobes, more or less obtuse at the apex, pubescent, to 6 cm long, to 3.5 cm wide, the upper leaves oblong to lanceolate, acute or subacute at the apex, simple or with 3 lobes, pubescent, often sessile; heads radiate, 1 to few on naked puberulent peduncles, 2.0–2.5 cm across, subtended by linear-lanceolate bractlets up to 7 mm long; involucre hemispheric; phyllaries in 2 series, the outer narrowly lanceolate, the inner lance-ovate, up to 10 mm long; ray flowers usually 8, yellow, 3- or 5-notched at the apex, pistillate, fertile, to 15 mm long; disc flowers up to 80, tubular, yellow, bisexual, fertile, the corolla 5-lobed, 4.5–5.5 mm long; cypselae oblong, narrowly winged, 2.5–3.0 mm long; pappus of 2 short scales or absent.

Common Name: Star tickseed.
Habitat: Dry soil.
Range: Virginia to Illinois to Nebraska, south to Texas and Florida.
Illinois Distribution: Confined to the southern fourth of the state.

The distinguishing features of this species are the general pubescence of its leaves and stems and its simple or 3-lobed leaves.

Coreopsis pubescens flowers from June to September.

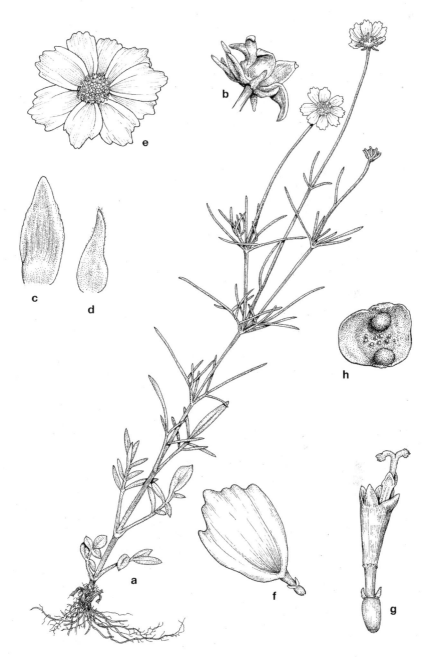

84b. *Coreopsis grandiflora*
var. *harveana*
(Narrow-leaved large-
flowered coreopsis).

a. Habit.
b. Cluster of phyllaries.
c. Outer phyllary.
d. Inner phyllary.

e. Flowering head.
f. Ray flower.
g. Disc flower.
h. Cypsela.

85. *Coreopsis pubescens*
   (Star tickseed).
a. Upper part of plant.

b. Outer phyllary.
c. Inner phyllary.
d. Ray flower.

e. Disc flower.
f. Cypsela.

8. **Coreopsis tinctoria** Nutt. Journ. Acad. Phila. 2:114. 1821. Fig. 86.

Annual herbs with a taproot; stems erect, branched, to 1.2 m tall, glabrous; leaves opposite, the lower 1 to 3 pinnately divided with segments linear to linear-lanceolate, the upper often undivided, glabrous, to 6 cm long, to 2.5 cm wide; heads radiate, 1 to several, on naked glabrous peduncles, 1.5–2.5 cm across, subtended by narrowly lanceolate bractlets up to 5 mm long; involucre hemispheric; phyllaries in 2 series, the outer narrowly oblong, the inner ovate, scarious-margined, to 6 mm long; receptacle flat, paleate, the paleae filiform, 4–5 mm long; ray flowers usually 8, yellow with a reddish brown blotch near the base, 3-notched at the apex, to 15 mm long, pistillate, fertile; disc flowers up to 80, tubular, reddish brown, bisexual, fertile, the corolla 4-lobed, 2.5–3.5 mm long; cypselae linear-oblong, very narrowly winged or wingless, 2–4 mm long; pappus absent or of 2 subulate scales.

Common Name: Golden coreopsis.
Habitat: Disturbed soil.
Range: Native to the western United States; often escaped from gardens.
Illinois Distribution: Scattered throughout Illinois.

This species is recognized by its divided leaves, its ray flowers that are yellow with a reddish brown basal blotch, and its 4-lobed, reddish brown disc corollas.

It is a fairly common garden escape into disturbed soil.

Coreopsis tinctoria flowers from July to September.

9. **Coreopsis verticillata** Sesse Moc. Pl. Nov. Hisp. 147. 1887. Not illustrated.

Perennial herbs with rhizomes; stems ascending to erect, slender, to 8 cm tall, sparsely branched, glabrous; leaves usually opposite, except sometimes for the uppermost, sessile, appearing verticillate, palmately compound, with 3 leaflets, the leaflets filiform, to 4 cm long, to 2 mm wide, glabrous, entire, acute at the apex; flowering heads radiate, few in a corymb, 2–4 cm across, on slender peduncles up to 4 cm long, with 5–6 bractlets at the base up to 5 mm long; involucre hemispheric; phyllaries 8, in two equal series, linear, usually glabrous, obtuse to acute; receptacle flat, paleate; ray flowers 8 per head, deep yellow, obtuse, not notched at the apex but sometimes erose, to 2 cm long, pistillate, fertile; disc flowers up to 80, tubular, yellow, bisexual, fertile, the corolla 5-lobed, 5–6 mm long; cypselae flat, oblong to obovate, 5–7 mm long, narrowly winged, glabrous or nearly so; pappus absent.

Common Name: Threadleaf coreopsis.
Habitat: Old field (in Illinois).
Range: Native to the southeastern United States and Mexico; often planted but rarely escaped from cultivation.
Illinois Distribution: DuPage County, collected by Scott Kobal in 2015.

This rarely escaped ornamental has the narrowest leaflets of any *Coreopsis* in Illinois.

This species flowers during July and August.

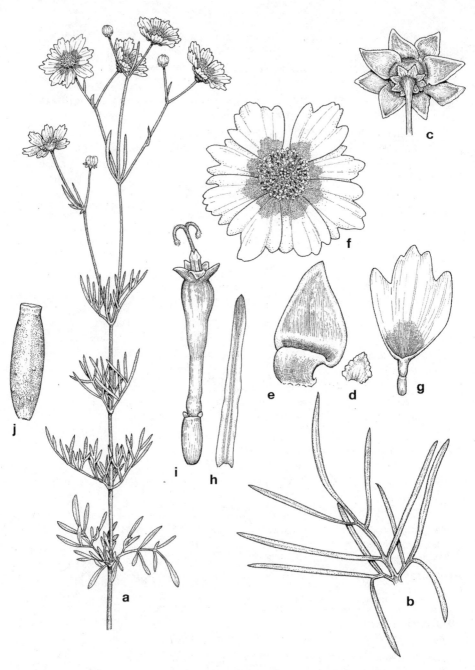

86. *Coreopsis tinctoria*
    (Golden coreopsis).
a. Upper part of plant.
b. Leaf.

c. Cluster of phyllaries.
d. Inner phyllary.
e. Outer phyllary.
f. Flowering head.

g. Ray flower.
h. Palea.
i. Disc flower.
j. Cypsela.

55. **Thelesperma** Less.—Greenthread

Annual or perennial herbs; stems erect, branched; leaves cauline, opposite, pinnately lobed; heads radiate or discoid, 1 to few in a corymb, subtended by three or more bractlets; involucre hemispheric to urceolate; phyllaries 5–8 in two series, equal, at least the inner connate for about half their length; receptacle flat or convex, paleate; ray flowers absent (in Illinois); disc flowers up to 100, tubular, yellow, bisexual, fertile, the corolla 5-lobed; cypselae usually terete, glabrous or verrucose; pappus of 2 retrorsely ciliate scales (in Illinois), sometimes absent.

This genus consists of ten species, all in the New World. It differs from *Coreopsis, Cosmos,* and *Bidens* by its terete cypselae and its inner phyllaries that are connate for about half their length.

Only the following species has been found in Illinois.

1. **Thelesperma megapotamicum** (Sprengel) Kuntze, Rev. Gen. Pl. 3:182. 1898. Fig. 87.
Bidens megapotamica Sprengel, Syst. Veg. 3:454. 1826.
Bidens gracilis Torr. Ann. Lyc. N.Y. 2:215. 1827.
Thelesperma gracile (Torr.) Gray, Kew Journ. Bot. 1:25. 1849.

Perennial herbs with a taproot; stems erect, branched, to 90 cm tall, glabrous; leaves cauline, opposite, 1- or 2-pinnatifid, its segments filiform to linear, to 4 cm long, to 2 mm wide, glabrous; heads discoid, 1 to few in a corymb, long-pedunculate, subtended by 3–8 ovate to oblong bractlets up to 2.5 mm long; involucre more or less hemispheric; phyllaries 5–8, in 2 series, the outer linear and spreading, the inner united for one-half their length, glabrous, with scarious margins; receptacle flat, paleate; ray flowers absent; disc flowers up to 100, tubular, yellow, bisexual, fertile, the corolla 5-lobed; cypselae terete, 5–8 mm long, glabrous; pappus of 2 retrorsely pubescent scales up to 2 mm long.

Common Name: Rayless thelesperma; greenthread.
Habitat: Disturbed soil.
Range: South Dakota to Utah, south to New Mexico, Texas, and Arkansas; adventive in Illinois and a few other states.
Illinois Distribution: Known only from Kane County.

Our specimen is an escape from cultivation. It is native in desert scrub and oak-juniper woodlands in the western United States.

Thelesperma megapotamicum flowers from June to October.

56. **Cosmos** Cav.—Cosmos

Annual or perennial herbs; stems usually erect, branched; leaves cauline, opposite, 1- to 3-pinnately divided, the segments usually entire; heads radiate, single or several in corymbs, subtended by up to 8 basally connate bractlets; involucre more or less hemispheric; phyllaries in 2 series, more or less equal, usually with a scarious margin; receptacle flat, paleate; ray flowers usually up to 8, variously colored,

87. *Thelesperma megapotanicum*
   (Rayless thelesperma).
a. Upper part of plant.
b. Disc flower.
c. Large inner phyllaries subtended
   by smaller outer phyllaries.

neutral; disc flowers up to 8, usually yellow, bisexual, perfect, the corolla 5-lobed; cypselae linear, usually 5-angled, short-beaked, usually wingless, glabrous, hispid, or papillose; pappus of 2–4 persistent awns.

Approximately twenty-five species are in the genus, all native to tropical and subtropical America. Our two species are both escapes from cultivation.

The genus is similar to *Bidens* because of its opposite leaves and pappus with 2–4 persistent awns, but differs by its 5-angled cypselae, its often pink, red, or purple flowers, and its fewer disc flowers per head. Its pubescent filaments of its stamens also distinguish it from *Bidens*, which has glabrous filaments.

1. Rays pink, red, or white; leaf segments linear to filiform . . . . 1. *C. bipinnatus*
1. Rays orange or golden yellow; leaf segments lanceolate to elliptic. . . . . 2. *C. sulphureus*

1. Cosmos bipinnatus Cav. Icon. 1:10, plate 14. 1791. Fig. 88.

Annual herbs with fibrous roots; stems ascending to erect, branched, to 2 m tall, glabrous or slightly pubescent; leaves cauline, opposite, much divided, to 10 cm long, glabrous or slightly pubescent, the segments linear to filiform, to 1.2 mm wide, sessile or on short petioles; heads radiate, several to numerous, on mostly naked peduncles up to 20 cm long, 1.0–1.5 cm across, subtended by several linear to linear-lanceolate bractlets up to 13 mm long; involucre hemispheric; phyllaries in 2 series, the outer lanceolate, the inner lance-ovate, to 13 mm long; receptacle flat, paleate; ray flowers usually 8, pink, red, or white, obovate to oblanceolate, shallowly notched at the apex, to 4 cm long, neutral; disc flowers up to 8, tubular, yellow, bisexual, fertile, the corolla 5-lobed, 5–7 mm long; cypselae more or less 5-angled, tapering to a beak, up to 15 mm long, glabrous, papillose; pappus absent or of 1–3 awns up to 3 mm long.

Common Name: Common cosmos.
Habitat: Disturbed areas.
Range: Native to Mexico; adventive in most of the United States.
Illinois Distribution: Scattered in the state.

This garden ornamental differs from *C. sulphureus* by its narrower leaf segments, smaller flowering heads, shorter cypselae, and its pink, red, or white rays.

Cosmos bipinnatus flowers from August to October.

2. Cosmos sulphureus Cav. Icon. 1:56, plate 79. 1791. Fig. 89.

Annual herbs with fibrous roots; stems erect, branched, up to 2 m tall, glabrous or sparsely hispid, ciliate; leaves cauline, opposite, much divided, to 15 cm long, glabrous or sparsely hispid, the segments lanceolate to elliptic, 2–5 mm wide, petiolate; heads radiate, several to numerous, 2–3 cm across, on mostly naked peduncles up to 20 cm long, subtended by several somewhat spreading, linear bractlets up to 8 (–10) mm long; involucre hemispheric; phyllaries in 2 series, lanceolate to oblong, to 15 mm long, usually with a scarious margin; ray flowers usually 8, orange to bright golden yellow, obovate, to 4 cm long, shallowly toothed at the apex, neutral; disc flowers up to 8, tubular, yellow, bisexual, fertile, the corolla 5-lobed,

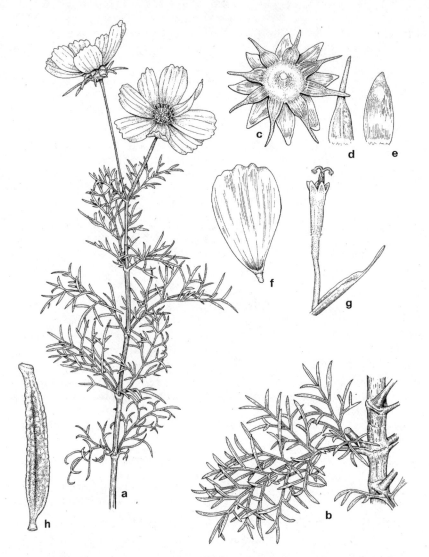

88. *Cosmos bipinnatus* (Common cosmos).
a. Upper part of plant.
b. Stem with nodes and lead.
c. Cluster of phyllaries.
d. Outer phyllary.
e. Inner phyllary.
f. Ray flower.
g. Disc flower with palea.
h. Cypsela.

5–7 mm long; cypselae more or less 5-angled, tapering to a beak, 15–30 mm long, usually hispidulous; pappus absent or of 2–3 awns up to 7 mm long.

Common Name: Yellow cosmos.
Habitat: Disturbed soil.
Range: Native to Mexico; adventive mostly in the eastern half of the United States.
Illinois Distribution: Known only from Kane County.

89. *Cosmus sulphureus*
(Yellow cosmos).
a. Upper part of
   plant.
b. Node with leaf.

c. Cluster of phyllaries.
d. Outer phyllary.
e. Inner phyllary.
f. Flowering head.
g. Ray flower.

h. Palea.
i. Disc flower.
j. Cypsela.
k. Cluster of cypselae.

The wider leaf segments, larger flowering heads, golden yellow or orange rays, and longer hispidulous cypselae distinguish this species from *C. bipinnatus*.

Cosmos sulphureus is a handsome garden ornamental. It flowers from August to October.

57. **Bidens** L.—Beggar-ticks; Bur Marigold.

Annual or perennial herbs; stems erect, occasionally angular, usually glabrous, usually with a pink or purplish spot at the nodes; leaves cauline, opposite, or the uppermost sometimes alternate, simple and unlobed or lobed or pinnately compound with up to 13 leaflets or sometimes bipinnate, usually glabrous or occasionally pilose or scabrous, petiolate or sessile; heads radiate or discoid, 1 to several, subtended by up to 20 erect, spreading, or reflexed bractlets; involucre hemispheric or campanulate; phyllaries in 2 series, the outer sometimes foliaceous, usually with a scarious margin; receptacle flat or somewhat convex, paleate; ray flowers when present 3–8, yellow or rarely white, neutral; disc flowers 12-numerous, tubular, yellow, bisexual, fertile, the corolla 4- or 5-lobed; cypselae flat or 3- or 4-angled, often ciliate, beakless, glabrous or pubescent; pappus usually of 2–4 ciliate or barbellate stiff awns.

As many as 250 species may be in this genus, widespread, particularly in North America and South America. One aquatic species considered by some to be in the genus *Bidens* I have elected to keep in a segregate genus *Megalodonta*.

1. Rays white . . . . .2. *B. alba*
1. Rays yellow, or absent.
  2. Leaves simple and unlobed or pinnately 3-lobed, never completely pinnately compound.
    3. All leaves pinnately 3-parted . . . . .11. *B. tripartita*
    3. Some or all leaves unlobed or divided, but not all leaves 3-parted.
      4. Leaves sessile or connate; heads nodding in fruit.
        5. Rays up to 15 mm long; outer cypselae 3–6 mm long, inner cypselae 4–8 mm long . . . . .9. *B. cernua*
        5. Rays (10-) 15–25 mm long; outer cypselae 6–8 mm long, inner cypselae 8–10 mm long . . . . .10. *B. laevis*
      4. Leaves petiolate or winged to the base; heads erect in fruit.
        6. Cypselae usually with 4 awns and tuberculate on the midvein; lobes of disc corollas 5. . . . . .13. *B. connata*
        6. Cypselae usually with 3 awns and smooth on the midvein; lobes of disc corollas 4 . . . . 12. *B. comosa*
  2. Most or all the leaves pinnately compound.
    7. Leaves bipinnate; cypselae linear, at least 8 times longer than broad . . . .
       . . . . .1. *B. bipinnata*
    7. Leaves once-pinnate; cypselae ovate to oblong, much less than 8 times longer than broad.
      8. Rays longer than the involucre; leaflets 3, 5, or 7, deeply cleft or coarsely serrate.
        9. Cypselae 2½–4 times as long as wide, with ciliate margins; leaflets common 7 . . . . 6. *B. trichosperma*
        9. Cypselae up to 2½ times as long as wide, with scabrous, rarely ciliate margins; leaflets commonly 3 or 5.

10. Outer phyllaries 8–12; leaflets usually 5- and 7-parted. . . . .7. *B. aristosa*
10. Outer phyllaries more than 12; leaflets usually 7- and 9-parted . . . .
. . . . 8. *B. polylepis*
8. Rays shorter than the involucre or usually absent; leaflets 3 or 5, uncleft.
11. Outer phyllaries 2–5, not ciliate . . . . 3. *B. discoidea*
11. Outer phyllaries 6 or more, ciliate.
12. Outer phyllaries usually 6–10; leaflets usually 5- or 3-parted . . . .
. . . . 4. *B. frondosa*
12. Outer phyllaries more than 10; leaflets usually 5-, 7-, or 9-parted . . . . . .
. . . . 5. *B. vulgata*

1. **Bidens bipinnata** L. Sp. Pl. 2:832. 1753. Fig. 90.

Annual herbs with fibrous roots; stems erect, branched, to 1.25 m tall, glabrous or puberulent, usually 4-angled; leaves cauline, opposite, 2- to 3-pinnately compound, to 20 cm long, the segments oblong, toothed or lobed, glabrous or hirtellous, petiolate; heads disciform, few to several, on slender, stiff, naked peduncles to 5 (–8) cm long, subtended by up to 10 linear appressed bractlets up to 5 mm long, ciliate; involucre campanulate; phyllaries in 2 series, linear to lanceolate, to 6 mm long; receptacle flat, paleate; ray flowers usually absent or, if present, about 5 in number and 1–2 mm long, yellow or whitish, neutral; disc flowers up to 25, tubular, yellow or less commonly whitish, bisexual, fertile, the corolla 5-lobed, 2–3 mm long; cypselae linear, 4-angled, tapering to the sharp-pointed apex, red-brown, at least 8 times longer than broad, the outer up to 15 mm long, the inner 12–18 mm long, hispidulous; pappus of 2–4 stiff retrorsely barbed awns up to 4 mm long.

Common Name: Spanish needles.
Habitat: Disturbed soil, open woods.
Range: New Brunswick to Ontario to Nebraska, south to New Mexico and Florida.
Illinois Distribution: Occasional in the southern three-fourths of the state, rare or
absent elsewhere.

This is the only *Bidens* in Illinois that has leaves that are 2- or 3-pinnately divided and with cypselae at least eight times longer than wide. When the cypselae are mature, the awns are stiff and retrorsely barbed and readily penetrate socks and trousers, often causing mild but irritating discomfort.

Some specimens lack ray flowers while others have about five ray flowers only 1–2 mm long.

Bidens bipinnata flowers from July to October.

2. **Bidens alba** (L.) DC. Prodr. 5:605. 1836. Fig. 91.
Coreopsis alba L. Sp.Pl. 2:908. 1753.

Annual herbs with fibrous roots; stems ascending to erect, usually branched, glabrous, to 50 cm tall (in Illinois); leaves cauline, opposite, simple and ovate to lanceolate or pinnately 3- to 7-lobed, to 8 cm long, to 2 cm wide, the lobes serrate or entire, ciliate, pubescent, petiolate; heads radiate, borne singly or several in

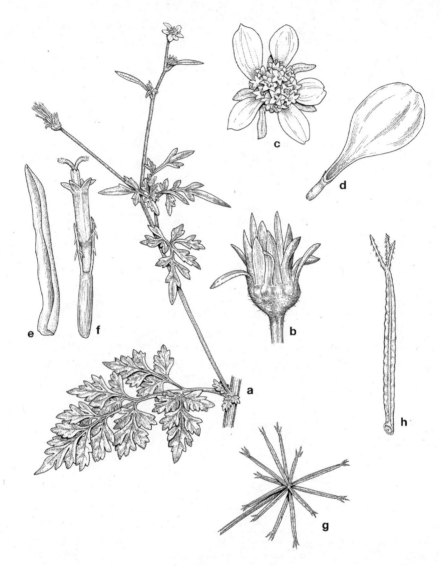

90. *Bidens bipinnata*
(Spanish needles).
a. Upper part of
plant.

b. Involucre of
phyllaries.
c. Flowering head.
d. Ray flower.

e. Palea.
f. Disc flower.
g. Cluster of cypselae.
h. Cypsela.

corymbs up to 2 cm across, on peduncles up to 3 (–5) cm long, subtended by up to
10 linear, appressed, ciliate, pubescent, the bractlets up to 5 mm long; involucre
campanulate; phyllaries in 2 series, the outer phyllaries 10–12 in number, lanceolate, up to 6 mm long; receptacle flat, paleate; ray flowers 5–8, white, up to 15 mm
long, neutral; disc flowers up to 40, tubular, yellow, bisexual, fertile, the corolla
5-lobed, up to 5 mm long; cypselae of two kinds, the outer red-brown, linear,

91. *Bidens alba* (White bidens).

a. Upper part of plant.
b. Cluster of phyllaries.
c. Outer phyllary.
d. Inner phyllary.
e. Disc flower.
f. Cypsela.

hispidulous, 3–5 mm long, the inner blackish, linear, 4-angled, hispidulous, 7–15 mm long; pappus of 2 awns 1–2 mm long.

Common Name: White bidens.
Habitat: Wet roadside ditch.
Range: Across the southern United States from Florida to California, extending north to Missouri, Illinois, and Pennsylvania.
Illinois Distribution: Known only from Massac County.

This is the only white-flowered *Bidens* in Illinois. Some botanists combine this species with *B. pilosa*, a view I do not agree with. I recognize three white-flowering species of *Bidens* in the southeastern United States. *Bidens pilosa* is usually rayless and has 3 awns per cypselae and usually 7–10 outer phyllaries; *Bidens odorata* has 5–8 rays per head and has 2 awns per cypselae or awnless and usually with about 8 outer phyllaries; *Bidens alba* has 5–8 rays per head and has 2 awns per cypselae and 10–12 outer phyllaries.

The only Illinois collection was made in a wet roadside ditch near Mermet in Massac County.

Bidens alba flowers during July and August in Illinois.

3. **Bidens discoidea** (Torr. & Gray) Britt. Bull. Torrey Club 20:281. 1893. Fig. 92.
Coreopsis discoidea Torr. & Gray, Fl. N. Am. 2:339. 1842.

Annual herbs with fibrous roots; stems erect, often much branched, to 1.5 cm tall, glabrous, often reddish, usually with purplish nodes; leaves cauline, opposite, the lower and middle ones pinnately 3-, 5-, or 7-parted, the upper often undivided, up to 8 cm long, up to 4 cm wide, glabrous or hirtellous, the leaflets lanceolate to ovate-lanceolate, serrate, ciliate, with short petiolules; heads discoid, few to several in corymbs, 3–10 mm across, on slender naked peduncles up to 5 cm long, subtended by 3 or 4 linear, appressed bractlets up to 2 mm long; involucre campanulate; phyllaries in 2 series, the outer 2–5 in number, ascending or spreading, narrowly oblong, glabrous, without cilia, to 10 mm long, the inner oblong, up to 6 mm long; receptacle flat, paleate; ray flowers absent or up to 4 mm long; disc flowers up to 20, tubular, orange, bisexual, fertile, the corolla 5-lobed, 1.5–2.0 mm long; cypselae linear to obovate to oblanceolate, more or less flattened, 3–6 mm long, usually not ciliate, short-hairy; pappus of 2 stiff, antrorsely pubescent awns 1.0–2.5 mm long.

Common Name: Swamp beggar-ticks; sticktight.
Habitat: Swamps, often on submerged logs.
Range: New Brunswick to Minnesota, south to Texas and Florida.
Illinois Distribution: Occasional in the southern two-fifths of Illinois; also Cook and DuPage counties.

Bidens discoidea is distinguished by its middle and lower leaves that are divided into three, five, or seven leaflets and the uppermost leaves undivided, the usual absence of ray flowers, and the 2–5 outer phyllaries that are glabrous.

This species is found mostly in southern Illinois where it often grows on logs that have been submerged in a swamp.

Bidens discoidea flowers from August to October.

4. **Bidens frondosa** L. Sp. Pl. 2:32. 1753. Fig. 93.
Bidens melanocarpa Wieg. Bull. Torrey Club 26:405. 1899.
Bidens melanocarpa Wieg. var. *pallida* Wieg. Bull. Torrey Club 26:406. 1899.

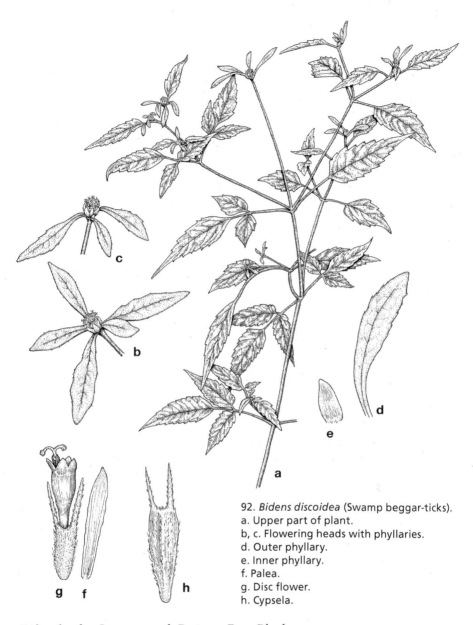

92. *Bidens discoidea* (Swamp beggar-ticks).
a. Upper part of plant.
b, c. Flowering heads with phyllaries.
d. Outer phyllary.
e. Inner phyllary.
f. Palea.
g. Disc flower.
h. Cypsela.

Bidens frondosa L. var. *anomala* Porter ex Fern. Rhodora 5:91. 1903.
Bidens frondosa L. f. *anomala* (Porter ex Fern.) Fern. Rhodora 40:352. 1938.
Bidens frondosa L. var. *pallida* (Wieg.) Wieg. Rhodora 26:5. 1924.

Annual herbs with fibrous roots; stems erect, branched, to 1 m tall, glabrous or nearly so, usually with purplish nodes; leaves cauline, opposite, the lower and middle ones pinnately 3- or 5-parted, the uppermost simple and undivided, to 10 cm long,

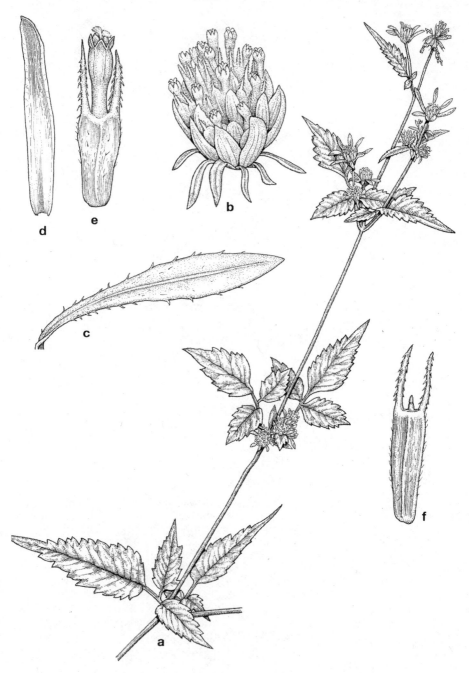

93. *Bidens frondosa*
(Common
beggar-ticks).
a. Upper part of plant.

b. Flowering head with
disc flowers.
c. Phyllary.

d. Palea.
e. Disc flower.
f. Cypsela.

to 6 cm wide, glabrous or hirtellous or ciliate, the leaflets lanceolate to lance-ovate, acuminate at the apex, serrate, with short petiolules; heads discoid, few to many in corymbs, 5–10 mm across, on naked peduncles up to 10 cm long, subtended by up to 10 foliaceous bractlets up to 3 cm long; involucre hemispheric to campanulate; phyllaries in 2 series, the outer 6 or more, ascending, narrowly oblong, ciliate, to 10 mm long, the inner oblong, to 6 mm long; receptacle flat, paleate; ray flowers absent or up to 4 mm long; disc flowers up to 100, tubular, yellow or orange-yellow, bisexual, fertile, the corolla 5-lobed, 2–3 mm long; cypselae usually nearly black, ciliate, the outer obovate, 5–7 mm long, the inner narrowly oblong, 7–10 mm long; pappus of 2 retrorsely or antrorsely pubescent awns up to 4 mm long.

Common Name: Common beggar-ticks; sticktights.
Habitat: Marshes, swamps, wet prairies, wet disturbed soil.
Range: Probably in every state; Mexico; Canada.
Illinois Distribution: Common throughout the state.

This species differs from *B. discoidea* by its ciliate bractlets 6–10 in number, and from *B. vulgata* by its bractlets not more than 10 in number.

Two forms occur in Illinois. This more common form has retrorsely barbed pappus awns and is the typical form. Plants with antrorsely barbed pappus awns are sometimes found and may be called f. *anomala*.

Bidens frondosa flowers from June to October.

5. **Bidens vulgata** Greene, Pittonia 4:72. 1899. Fig. 94.
Bidens frondosa L. var. *puberula* Wieg. Bull. Torrey Club 26:408. 1899.
Bidens vulgata Greene var. *puberula* (Wieg.) Greene, Pittonia 4:250. 1901.

Annual herbs with fibrous roots; stems erect, branched, to 1.2 m tall, glabrous or puberulent, usually with purplish nodes; leaves cauline, opposite, the lower and middle ones 5-, 7-, or 9-parted, to 12 cm long, to 10 mm wide, the uppermost usually undivided, the leaflets usually lanceolate, acute at the apex, dentate or serrate, glabrous or hispidulous; heads discoid, 1 to few in a corymb, 5–10 mm across, on naked peduncles up to 15 cm long, subtended by 15–20 linear, ciliate bractlets up to 4 cm long; involucre campanulate to hemispheric; phyllaries 10 or more, in 2 series, oblong to narrowly ovate, to 10 mm long; receptacle flat, paleate; ray flowers absent or up to 4 mm long; disc flowers up to 100, tubular, yellow, bisexual, fertile, the corolla 5-lobed, 2.5–3.5 mm long; cypselae usually purple or brown, obovate, ciliate, flattened, the outer 6–10 mm long, the inner 8–15 mm long; pappus of 2 awns, those of the outer cypselae 3–6 mm long, those of the inner cypselae 4–10 mm long, the awns retrorsely or antrorsely barbed.

Common Name: Tall beggar-ticks; tall sticktights.
Habitat: Moist, often disturbed soil.
Range: Throughout most of the United States; Mexico; Canada.
Illinois Distribution: Occasional throughout the state.

94. *Bidens vulgata*
(Tall beggar-
ticks).
a. Upper part of
plant.

b. Flowering head
subtended by large
phyllaries.
c. Outer phyllary.
d. Inner phyllary.

e. Palea.
f. Ray flower.
g. Disc flower.
h. Cypsela.

This species is similar to *B. discoidea* and *B. frondosa* by its pinnately compound middle and lower leaves with 5, 7, or 9 leaflets and its usually rayless flowering heads. It differs from *B. discoidea* by its ciliate phyllaries and from *B. frondosa* by having 10 or more phyllaries per flowering head.

Rarely are plants found with puberulent stems and leaves. These have been called var. *puberula*.

Bidens vulgata flowers from August to October.

6. **Bidens trichosperma** (Michx.) Britt. Bull. Torrey Club 20:281. 1893. Figs. 95a, 95b.

Coreopsis coronata L. Sp. Pl. 2:1281. 1753.
Coreopsis trichosperma Michx. Fl. Bor. Am. 2:139. 1803.
Bidens coronata (L.) Fisch. & Steud. Nomencl. Bot., ed. 2, 202. 1840, *non B. coronata*
Fisch. & Colla (1834).

Annual herbs with fibrous roots; stems erect, branched, to 1.5 m tall, usually glabrous, often with purplish nodes; leaves cauline, opposite, the lower and middle ones pinnately divided into 7 or 9 leaflets to 10 cm long, to 3 cm wide, the uppermost less commonly undivided, usually glabrous, the leaflets linear to narrowly lanceolate, acuminate at the apex, dentate to coarsely serrate, on petioles to 5 cm long; heads radiate, solitary to several in corymbs, 1.5–2.5 cm across, on naked peduncles up to 7 cm long, subtended by about 8 spreading to ascending, sometimes foliaceous bractlets up to 2.5 (–3.5) cm long; involucre hemispheric or campanulate; phyllaries in 2 series, the inner oblong to narrowly ovate, to 10 mm long; receptacle flat, paleate; ray flowers 8–12, yellow, 1.2–2.5 cm long, neutral; disc flowers up to 100, tubular, yellow, bisexual, fertile, the corolla 5-lobed, 2.5–3.0 mm long; cypselae more or less flat, ciliate, brown to black, the outer 3–6 mm long, the inner 5–9 mm long, usually glabrous; pappus with 2 retrorsely or antrorsely barbed awns, the awns of the outer cypselae 0.5–2.5 mm long, the awns of the inner cypselae 0.5–3.5 mm long.

This species has usually been called *B. coronata* in the past, but Britton's use of this binomial is preempted by Colla's use of the same binomial for a different species.

Two varieties occur in Illinois:
a. Some or all the inner cypselae 6 mm long or longer; leaflets narrowly lanceolate, most of them more than 12 mm wide .... 6a. *B. trichosperma* var. *trichosperma*
a. None of the inner cypselae more than 6 mm long; leaflets linear, up to 12 mm wide ...
....6b. *B. trichosperma* var. *tenuiloba*

6a. **Bidens trichosperma** (Michx.) Britt. var. **trichosperma** Fig. 95a.
Some or all the inner cypselae 6 mm long or longer; leaflets narrowly lanceolate, most of them more than 12 mm wide.

Common Name: Tall swamp marigold; sticktights.
Habitat: Wet ground.
Range: Quebec to Minnesota to South Dakota, south to Nebraska, Arkansas, Louisiana, and Florida.
Illinois Distribution: Scattered throughout Illinois.

This variety differs by its longer cypselae and its broader leaflets, but there is intergradation with the next variety.
This is one of the showiest species of *Bidens* in Illinois.
Bidens trichosperma var. *trichosperma* flowers from June to October.

95a. *Bidens trichosperma*
    var. *trichosperma* (Tall
    swamp marigold).

a. Upper part of plant.
b. Outer phyllary.
c. Inner phyllary.

d. Ray flower.
e. Disc flower.
f. Cypsela.

6b. **Bidens trichosperma** (Michx.) Britt. var. **tenuiloba** (Gray) Britt. Bull. Torrey Club 20:281. 1893. Fig. 95b.

Coreopsis trichosperma Michx. var. *tenuiloba* Gray, Syn. Fl. N. Am. 1:295. 1884.
Bidens coronata (L.) Britt. var. *tenuiloba* (Gray) Sherff, Bot. Gaz. 86:446. 1928.

None of the cypselae more than 6 mm long; leaflets linear, up to 12 mm wide.

Common Name: Tall swamp marigold; sticktights.
Habitat: Wet ground.
Range: Ontario to Minnesota to Nebraska, south to Missouri, Tennessee, and South Carolina.
Illinois Distribution: Common in the northern two-thirds of Illinois, occasional elsewhere.

This variety appears to be the more common one in the northern half of the state. It flowers from June to October.

7. **Bidens aristosa** (Michx.) Britt. Bull. Torrey Club 20:281. 1893. Fig. 96.
Coreopsis aristosa Michx. Fl. Bor. Am. 2:140. 1803.

Annual herbs with fibrous roots; stems erect, branched, to 1 m tall, glabrous, usually purplish at the nodes; leaves cauline, opposite, the lower and middle divided into 5 or 7 leaflets, to 10 cm long, to 4 cm wide, usually glabrous, the leaflets lanceolate, acuminate at the apex, ciliate, serrate; heads radiate, few to several in corymbs, 2–3 cm across, subtended by 5–10 nonfoliaceous, linear, usually glabrous, up to 10 mm long; involucre hemispheric; phyllaries in 2 series, the outer narrowly ovate, the inner broadly linear, to 8 (–10) mm long; receptacle flat, paleate; ray flowers up to 10, yellow, to 2.5 cm long, neutral; disc flowers up to 40 (–50), tubular, yellow, bisexual, fertile, the corolla 5-lobed, 2–3 mm long; cypselae flat, brown to black, narrowly winged, glabrous or tuberculate, the outer 4–7 mm long, the inner 5–8 mm long; pappus of 2 antrorsely or retrorsely barbed awns up to 5 mm long, or absent.

Bidens aristosa is similar to *B. trichosperma* and *B. polylepis* by its pinnately compound leaves and its attractive radiate flowering heads. It differs from *B. trichosperma* by its broader leaflets and non-ciliate cypselae. It differs from *B. polylepis* by its fewer bractlets subtending the flowering heads.

Two forms and one variety occur in Illinois.
a. Cypselae awned.
  b. Cypselae with the awns antrorsely barbed. . . . . 7a. *B. aristosa* f. *aristosa*
  b. Cypselae with the awns retrorsely barbed . . . . 7b. *B. aristosa* f. *fritcheyi*
a. Cypselae without awns . . . . 7c. *B. aristosa* var. *mutica*

7a. **Bidens aristosa** (Michx.) Britt. var. **aristosa** f. **aristosa** Fig. 130 (c) (see p. 288). Cypselae with the awns antrorsely barbed.

Common Name: Swamp marigold.
Habitat: Wet ground.
Range: Ontario to Minnesota, south to Texas, Alabama, and South Carolina.

95b. *Bidens trichosperma*
var. *tenuiloba* (Tall
swamp marigold).
a. Upper part of plant.

a

96. *Bidens aristosa*
var. *mutica*
(Swamp
marigold).

a. Upper part of plant.
b. Outer and inner
   phyllaries.
c. Flowering head.

d. Ray flower.
e. Disc flower with palea.
f. Disc flower.
g. Cypsela.

Illinois Distribution: Common in the southern two-thirds of Illinois, less common elsewhere.

This form seems to be a little more common than f. *fritcheyi*.
The typical form flowers from August to October.

7b. **Bidens aristosa** (Michx.) Britt. var. **aristosa** f. **fritcheyi** (Fern.) Wunderlin, Ann. Mo. Bot. Gard. 59:471. 1972. Fig. 130 (d) (see p. 288).
Bidens aristosa (Michx.) Britt. var. *fritcheyi* Fern. Rhodora 15:78. 1913.

Cypselae with the awns retrorsely barbed.

Common Name: Swamp marigold.
Habitat: Wet ground.
Range: Mostly in the central midwestern United States.
Illinois Distribution: Occasional and scattered throughout the state.

Since the two types of awns are similar to those in *B. frondosa*, it seems logical to treat these consistently as forms.

This form flowers from August to October.

7c. **Bidens aristosa** (Michx.) Britt. var. **mutica** (Gray) Gattinger ex Fern. Rhodora 15:78. 1913.
Coreopsis aristosa (Michx.) Britt. var. mutica Gray, Man. Bot. N. U.S., ed. 5, 260. 1867.
Bidens aristosa (Michx.) Britt. f. *mutica* (Gray) Wunderlin, Ann. Mo. Bot. Gard. 59:471. 1972.

Cypselae without awns.

Common Name: Swamp marigold.
Habitat: Wet ground.
Range: Similar to f. *fritcheyi*.
Illinois Distribution: Rare but scattered in Illinois.

The absence of awns on the cypselae clearly distinguishes this variety.

It flowers from August to October.

8. **Bidens polylepis** S.F. Blake, Proc. Biol. Sci. Wash. 35:78. 1922. Figs. 97, 103 (b).
Coreopsis involucrata Nutt. Journ. Acad. Nat. Sci. Phila. 7:74. 1834.
Bidens involucrata (Nutt.) Britt. Bull. Torrey Club 20:28. 1893, *non* Philippi (1891).
Bidens aristosa (Michx.) Britt. f. *involucrata* (Nutt.) Wunderlin, Ann. Mo. Bot. Gard. 59:472. 1972.

Annual herbs with fibrous roots; stems erect, branched, up to 1.2 m tall, glabrous or nearly so, usually purplish at the nodes; leaves cauline, opposite, the lower and middle ones pinnately divided into (5) 7 or 9 leaflets, to 10 cm long, 2–6 cm wide, glabrous or nearly so, ciliate, the leaflets linear to lanceolate, acuminate at the apex, serrate or dentate; heads radiate, few to several in corymbs, 2–3 cm across, on naked peduncles up to 3 (–4) cm long, subtended by 12 or more linear, hispid, ciliate, nonfoliaceous bractlets to 12 (–15) mm long; involucre hemispheric; phyllaries in 2 series, lanceolate to narrowly ovate, up to 8 mm long; receptacle flat, paleate; ray flowers usually 8, yellow, up to 2.5 cm long, neutral; disc flowers up to 75, tubular, yellow, bisexual, fertile, the corolla 5-lobed, 2.5–3.0 mm long; cypselae brown or black, flattened, oblanceolate, hispid, ciliate, the outer 4.5–5.5 mm long, the inner 5.5–6.5 mm long; pappus of 2 antrorsely or retrorsely barbed awns up to 1 mm long.

The most obvious differences between this species and *B. aristosa* are the twelve or more bractlets that subtend each flowering head and the much shorter awns of the cypselae. Recent botanists have considered this species to be the same as *B. aristosa*, a view I do not share.

Two forms occur in Illinois.

a. Awns of cypselae antrorsely barbed . . . . 8a. *B. polylepis* f. *polylepis*
a. Awns of cypselae retrorsely barbed. . . . . 8b. *B. polylepis* f. *retrorsa*

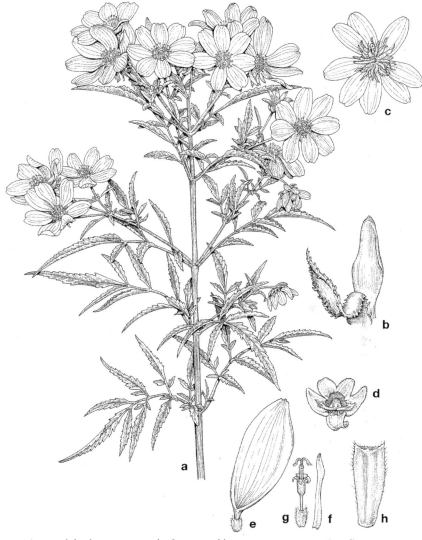

97. *Bidens polylepis* (Many-bracted bur marigold).
a. Upper part of plant.

b. Outer and inner phyllaries.
c. Flowering head.
d. Disc flower.

e. Ray flower.
f. Palea.
g. Disc flower.
h. Cypsela.

8a. **Bidens polylepis** S.F. Blake f. **polylepis**
Awns of cypselae antrorsely barbed.

Common Name: Many-bracted bur marigold.
Habitat: Marshes, swamps, wet ditches.
Range: Pennsylvania to Wisconsin and Nebraska and Colorado, south to New
    Mexico, Texas, Alabama, and South Carolina.
Illinois Distribution: Occasional to common throughout Illinois.

This is one of the most attractive species of *Bidens* when in flower. The typical form
is the more common of the two forms in Illinois.

This form flowers from August to October.

8b. **Bidens polylepis** S.F. Blake f. **retrorsa** (Sherff) Mohlenbr. **comb. nov.**
    (Basionym: *Bidens polylepis* S.F. Blake var. *retrorsa* Sherff, Bot. Gaz. 80:386.
    1925.) Fig. 130 (b) (see p. 288).
Bidens polylepis S.F. Blake var. *retrorsa* Sherff, Bot. Gaz. 80:386. 1925.
Bidens aristosa (Michx.) Britt. f. *retrorsa* (Sherff) Wundelin, Ann. Mo. Bot. Gard.
    59:471. 1972.

Awns of cypselae retrorsely barbed.

Common Name: Many-bracted bur marigold.
Habitat: Wet ground.
Range: Ohio to Missouri.
Illinois Distribution: Mostly in the central counties of Illinois.

In order to be consistent, I am reducing the retrorsely awned plants to forms
instead of varieties.

This form flowers from August to October.

9. **Bidens cernua** L. Sp. Pl. 2:832. 1753. Figs. 98a, 98b, 98c.
Bidens filamentosa Rydb. Brittonia 1:104. 1931.

Annual herbs with fibrous roots, often decumbent and rooting at the nodes;
stems erect, branched, to 1 m tall, glabrous or usually somewhat hispid; leaves
cauline, opposite, simple, linear to lanceolate to narrowly ovate, acute to acuminate at the apex, to 8 cm long, to 2.5 cm wide, nearly entire to dentate, rounded
at the connate base or tapering to the sessile base, glabrous, sometimes ciliate;
heads radiate, one to a few in a corymb, 1.0–2.2 cm across, on naked peduncles up
to 4 (–6) cm long, nodding as it ages, subtended by up to 10 spreading or reflexed,
narrowly oblong, ciliate, usually foliaceous bractlets up to 2 cm long; involucre
hemispheric; phyllaries in 2 series, 6–8 (–10) in number, lanceolate to lance-ovate,
to 10 mm long; receptacle flat, paleate; ray flowers 6–8, yellow, to 15 mm long,
elliptic, neutral; disc flowers 50–100, tubular, yellow or yellow-orange, bisexual,
fertile, the corolla 5-lobed, 3–4 mm long; cypselae black or brown, somewhat

flattened, often 4-angled, retrorsely hispid, the outer 3–6 mm long, the inner 4–8 mm long; pappus of usually 4 retrorsely barbed awns 2.5–4.0 mm long.

This species is similar to *B. laevis*, but differs by its smaller rays and smaller cypselae. When the flowering heads open, they are erect, but as the head matures, the peduncle bends and the head becomes nodding.

Three varieties occur in Illinois.
a. Leaves rounded at the connate base.
    b. Leaves coarsely toothed, the teeth 1 mm long or usually longer ....
    ....9a. *B. cernua* var. *cernua*
    b. Leaves finely toothed, the teeth up to 1 mm long, or absent . . 9b. *B. cernua* var. *integra*
a. Leaves tapering to the sessile base .... 9c. *B. cernua* var. *elliptica*

98a. *Bidens cernua* var.
    cernua (Nodding bur
    marigold).
a. Upper part of plant.
b. Leaf.
c. Flowering head.

9a. **Bidens cernua** L. var. **cernua** Fig. 98a.

Leaves rounded at the connate base; leaves coarsely toothed, the teeth 1 mm long or usually longer.

Common Name: Nodding bur marigold.
Habitat: Wet ground, including bogs.
Range: Nova Scotia to British Columbia, south to California, South Dakota, and Illinois.
Illinois Distribution: Occasional throughout the state.

This is the most common variety in Illinois. It flowers from June to October.

9b. **Bidens cernua** L. var. **integra** Wieg. Bull. Torrey Club 26:418–419. 1899. Fig. 98b.

Leaves rounded at the connate base; teeth of the leaves up to 1 mm long, or absent.

Common Name: Nodding bur marigold.
Habitat: Shores of ponds and streams, bogs, marshes.
Range: Prince Edward Island to North Dakota, south to Oklahoma and North Carolina.
Illinois Distribution: Infrequently collected in most of Illinois.

This is seemingly the least common variety of *B. cernua* in Illinois. It flowers from June to October.

9c. **Bidens cernua** L. var. **elliptica** Wieg. Bull. Torrey Club 26:417–418. 1899. Fig. 98c.

Leaves tapering to the sessile base.

Common Name: Nodding bur marigold.
Habitat: Wet ground.
Range: Quebec to Washington, south to Kansas and North Carolina.
Illinois Distribution: Occasional in Illinois.

In this variety, the leaves are not connate below. It flowers from June to October.

10. **Bidens laevis** (L.) BSP. Prel. Cat. 29. 1888. Fig. 99.
Helianthus laevis L. Sp. Pl. 2:906. 1753.

Annual herbs with fibrous roots; stems erect, branched, to 1 m tall, glabrous, usually purplish at the nodes; leaves cauline, opposite, simple, undivided, to 10 cm long, to 3.5 cm wide, linear-lanceolate to elliptic, acuminate at the apex, tapering to the sessile and usually connate base, glabrous, serrate; heads radiate, 1 to several in a corymb, 2.5–4.0 cm across, on naked peduncles up to 6 cm long, subtended by up to 8 linear to lanceolate, ciliate, usually foliaceous bractlets up

98b. *Bidens cernua* var.
    integra (Nodding bur
    marigold).
d. Habit.

e. Node with opposite
    leaves.
f. Outer phyllary.
g. Inner phyllary.

h. Flowering head.
i. Disc flower.
j. Cypsela.

98c. *Bidens cernua* var.
  elliptica (Nodding bur
  marigold).
a. Upper part of plant.

b. Node.
c. Leaf.
d. Phyllaries.
e. Ray flower.

f. Disc flower with palea.
g. Disc flower with
  cypsela.
h. Cypsela.

to 1.5 (−2.0) cm long; involucre hemispheric to turbinate; phyllaries in 2 series, narrowly ovate to ovate, up to 8 mm long; receptacle flat, paleate; ray flowers yellow or golden yellow, oval to elliptic, usually 8 in number, (10−) 15–25 mm long, neutral; disc flowers up to 100, tubular, yellow or golden yellow, bisexual, fertile, the corolla 5-lobed, 4.0–6.5 mm long; cypselae usually black or red-brown, flat or sometimes angled, glabrous or ciliate, the outer 6–8 mm long, the inner 8–10 mm long; pappus of 2 or 4 retrorsely barbed awns 3–5 mm long.

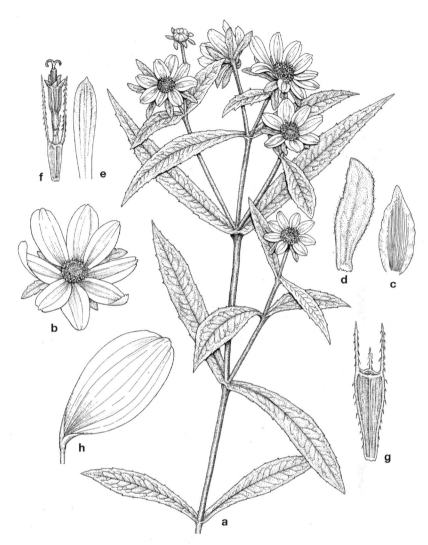

99. *Bidens laevis* (Large-flowered beggar-ticks).
a. Upper part of plant.
b. Flowering head. c,
d. Phyllaries.
e. Palea.
f. Disc flower.
g. Cypsela.
h. Ray flower.

Common Name: Large-flowered beggar-ticks.
Habitat: Wet soil.
Range: Maine to Missouri to California, south to Arizona, Texas, and Florida.
Illinois Distribution: Known only from Union County.

This *Bidens* has the largest and showiest flowering heads of any in Illinois. In fact, Linnaeus originally described it as a *Helianthus*, or sunflower. Although similar to *B. cernua*, it is larger in all respects.

Its single Illinois location is at the edge of LaRue Swamp in Union County. Pepoon's report of this species from Cook County is an error for *B. cernua*. *Bidens laevis* flowers in August and September.

11. **Bidens tripartita** L. Sp. Pl. 2:831. 1753. Fig. 100.
Annual herbs with fibrous roots; stems erect, branched, to 50 (–75) cm tall, glabrous, usually purplish at the nodes; leaves cauline, opposite, deeply 3-lobed or sometimes compound with 3 leaflets, and with no undivided leaves, to 8 cm long, to 4 cm wide, the lobes usually dentate, rarely entire, glabrous, ciliate, on petioles to 2 cm long; heads discoid, several in a corymb, 5–12 mm across, on naked peduncles up to 6 cm long, subtended by up to 10 linear-lanceolate, strongly hirsute, more or less foliaceous bractlets up to 3.5 cm long; involucre hemispheric; phyllaries in 2 series, narrowly ovate, usually ciliate, to 10 mm long; receptacle flat, paleate; ray flower absent; disc flowers up to 100, tubular, yellow, bisexual, fertile, the corolla 5-lobed, 3–4 mm long; cypselae black, flat, retrorsely ciliate, 6–8 mm long; pappus of 3 or 4 retrorsely barbed awns up to 3 (–5) mm long.

Common Name: Three-parted beggar-ticks.
Habitat: Wet ground.
Range: Throughout most of the United States.
Illinois Distribution: Scattered and fairly common throughout Illinois.

I am restricting *B. tripartita* to those plants whose every leaf is 3-parted, whereas the very similar *B. comosa* has at least some leaves undivided. Also, *B. tripartita* has more strongly hirsute bractlets subtending the flowering heads, and the cypselae and its awns are usually slightly shorter.
    Bidens tripartita flowers from July to October.

12. **Bidens comosa** (Gray) Wieg. Bull. Torrey Club 24:436. 1897. Fig. 101.
Bidens connata Muhl. ex Willd. var. *comosa* Gray, Man. Bot. N. U.S., ed. 5, 261. 1867.
Bidens comosa (Gray) Wieg. var. *acuta* Wieg. Bull. Torrey Club 26:411. 1899.
Annual herbs with fibrous roots; stems erect, branched, to 60 cm tall, glabrous, usually with purplish nodes; leaves cauline, opposite, mostly undivided or with some or all the lower and middle leaves 3-lobed, to 6 cm long, to 3 cm wide, dentate or serrate, glabrous, sessile or on petioles to 1 (–2) cm long; heads discoid, several in a corymb, 5–10 mm across, on naked peduncles up to 5 cm long, subtended by up to 10 linear-lanceolate, sparsely ciliate, sometimes foliaceous bractlets to 6 cm long; involucre campanulate; phyllaries in 2 series, ovate-lanceolate, sparsely ciliate, to 12 mm long; receptacle flat, paleate; ray flowers absent; disc flowers up to 60, tubular, yellow, bisexual, fertile, the corolla 4-lobed, 2.5–3.5 mm long; cypselae flat, oblanceolate, retrorsely ciliate, the outer 5–9 mm long, the inner 8.5–11.0 mm long; pappus of usually 3 retrorsely barbed awns up to 6 mm long.

Common Name: Swamp beggar-ticks; swamp tickseed.
Habitat: Wet ground, sometimes weedy.

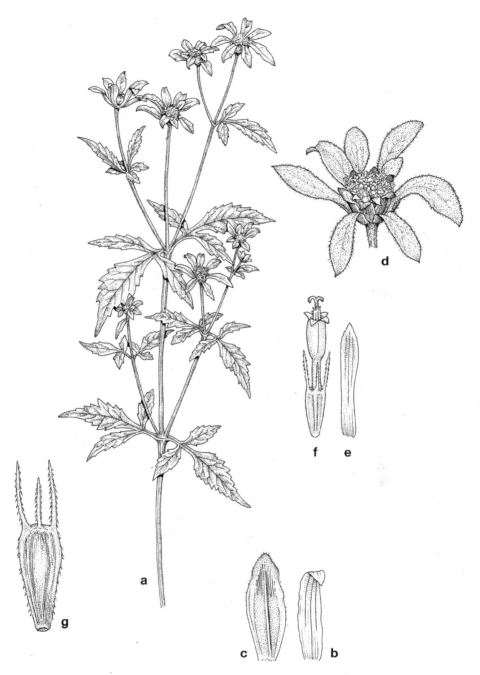

100. *Bidens tripartita*
(Three-parted
beggar-ticks).

a. Upper part of plant.
b. Outer phyllary.
c. Inner phyllary.
d. Flowering head.

e. Palea.
f. Disc flower.
g. Cypsela.

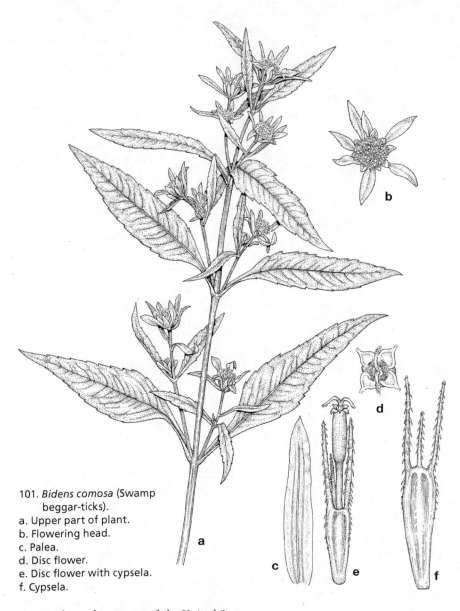

101. *Bidens comosa* (Swamp
beggar-ticks).
a. Upper part of plant.
b. Flowering head.
c. Palea.
d. Disc flower.
e. Disc flower with cypsela.
f. Cypsela.

Range: Throughout most of the United States.
Illinois Distribution: Common throughout Illinois.

This species is very similar to *B. connata*, but differs by its sessile leaves, its 4-lobed
disc corollas, and the three awns of the cypselae. The somewhat similar *B. tripartita* has all leaves 3-parted.

    Bidens comosa flowers during September and October.

13. **Bidens connata** Muhl. ex Willd. Sp. Pl. 3:1718. 1803. Fig. 102.

Annual herbs with fibrous roots; stems erect, branched, often purple, to 80 cm tall, glabrous; leaves cauline, opposite, unlobed or some of them 3- or 5-lobed, to 10 cm long, to 3.5 cm wide, acuminate at the apex, tapering to a petiolate base, glabrous, ciliate, serrate to dentate, rarely entire; heads discoid, 1 to several in a corymb, 5–12 mm across, on naked peduncles up to 5 (–7) cm long, subtended by up to 6 linear to oblong, ciliate, foliaceous bractlets up to 2.5 cm long; involucre hemispheric to campanulate; phyllaries in 2 series, up to 8 in number, narrowly ovate, to 10 mm long; receptacle flat, paleate; ray flowers absent; disc flowers up to 50, tubular, yellow, bisexual, fertile, the corolla 5-lobed, 2–3 mm long; cypselae black, flat or 4-angled, retrorsely ciliate, the outer 3–7 mm long, the inner 4–8 mm long; pappus of usually 4 retrorsely (in Illinois) barbed awns up to 5 mm long, or rarely with both retrorsely and antrorsely barbed awns on the same plant.

Two varieties occur in Illinois:

a. Cypsela with only retrorse bristles.
    b. Most but not all of the leaves 3- or 5-lobed. . . . . 13a. *B. connata* var. *connata*
    b. Most of the leaves undivided . . . . 13b. *B. connata* var. *petiolata*
a. Cypsela with both retrorse and antrorse bristles on same plant. . . . .
    . . . . 13c. *B. connata* var. *ambiversa*

13a. **Bidens connata** Muhl. ex Willd. var. **connata** Fig. 102 (a–f).
Most or not all of the leaves 3- or 5-lobed; cypselae with retrorse bristles only.

Common Name: Swamp beggar-ticks; pink-stemmed beggar-ticks.
Habitat: Wet ground, particularly in muddy areas after water has receded.
Range: Newfoundland to Minnesota to Montana, south to Kansas, Tennessee, and
    Georgia.
Illinois Distribution: Occasional throughout the state.

Bidens connata differs from *B. comosa* by its 5-lobed disc corollas and its usually 4-angled cypselae. It differs from *B. tripartita* by having at least some of the leaves undivided.

This plant flowers from August to October.

13b. **Bidens connata** Muhl. ex Willd. var. **petiolata** (Nutt.) Farw. Rep. Comm.
Parks Boul. Detroit 11:91. 1900. Fig. 102 (a–e).
Bidens petiolata Nutt. Journ. Acad. Nat. Sci. Phila. 7:99–100. 1834.

Most of the leaves undivided; cypselae with retrorse bristles only.

Common Name: Swamp beggar-ticks; pink-stemmed beggar-ticks.
Habitat: Wet ground.
Range: Same as var. *connata*.
Illinois Distribution: Occasional throughout the state.

This variety flowers from August to October.

102. *Bidens connata* var. *connata* (Common beggar-ticks).

a. Habit.
b. Outer phyllary.
c. Inner phyllary.
d. Disc flower.
e. Cypsela.
f. Three-lobed leaf.

13c. **Bidens connata** Muhl. ex Willd. var. **ambiversa** Fassett, Rhodora 30:33–34. 1928. Not illustrated.

Most of the leaves undivided; cypselae with both retrorse and antrorse bristles on the same plant.

Common Name: Swamp beggar-ticks.
Habitat: Wet ground.
Range: Same as var. *connata.*
Illinois Distribution: Rare in the state.

This variety flowers from August to October.

58. **Megalodonta** Greene—Water Marigold

Aquatic perennials; stems submersed, unbranched or branched, glabrous; leaves opposite or sometimes whorled, the submersed ones divided into filiform segments, the emersed ones deeply lobed to serrate; head radiate, solitary, on glabrous peduncles, subtended by 5–6 bractlets; involucre hemispheric; phyllaries in 2 series, the outer smaller than the inner; receptacle flat, paleate; ray flowers 8, yellow, showy, neutral; disc flowers up to 50, tubular, yellow, bisexual, fertile, the corolla 5-lobed; cypselae terete or slightly 4-angled, without cilia, glabrous; pappus of 2–6 retrorsely barbed awns as long as or longer than the cypselae.

Two or three species comprise this genus.

This genus is sometimes merged with *Bidens*, but the species of *Megalodonta* are aquatic perennials with filiform submersed leaflets and the cypselae have awns equaling or longer than the cypselae.

Only the following species occurs in Illinois.

1. **Megalodonta beckii** (Torr. ex Sprengel) Greene, Pittonia 4:271. 1901. Fig. 103. *Bidens beckii* Torr. ex Sprengel, Neue Entd. 2:135. 1821.

Aquatic perennials; stems submersed, branched or unbranched, to 2 m long, glabrous; leaves opposite or sometimes whorled, the submersed ones filiform, the emersed ones lance-ovate to ovate, acute at the apex, rounded at the sessile base, glabrous, deeply lobed or serrate; head radiate, solitary, showy, on peduncles up to 8 (–10) cm long, subtended by 5–6 oblong, ciliate bractlets up to 8 mm long; involucre hemispheric; phyllaries in 2 series, glabrous, the outer smaller than the inner, to 10 mm long; receptacle flat, paleate; ray flowers 8, yellow, up to 15 mm long, neutral; disc flowers up to 30, tubular, yellow, bisexual, fertile, the corolla 5-lobed, 5–6 mm long; cypselae terete or slightly 4-angled, glabrous, eciliate, 10–15 mm long; pappus of 2–6 retrorsely barbed awns up to 40 mm long, as long as or longer than the cypselae.

Common Name: Water marigold.
Habitat: Ponds and deep, clear lakes in standing water.

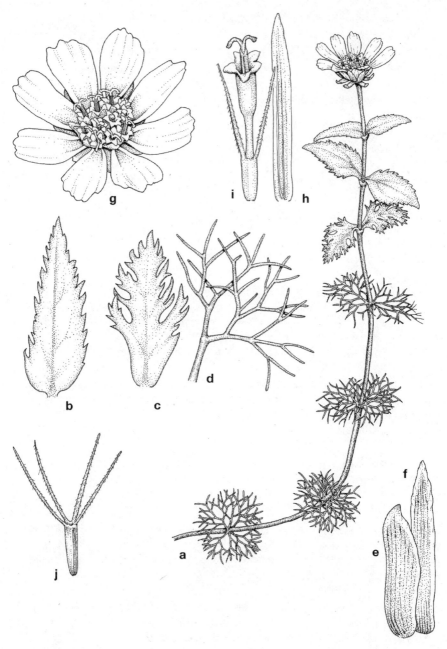

103. *Megalodonta beckii* (Water marigold).

a. Upper part of plant.

b, c. Variation of upper leaves.
d. Lower leaf.
e. Outer phyllary.
f. Inner phyllary.

g. Flowering head.
h. Palea.
i. Disc flower.
j. Cypsela.

Range: Maine to Minnesota, south to Missouri and Illinois.

Illinois Distribution: Very rare; known only from Cook, Lake, and St. Clair counties.

This is a very distinctive aquatic plant with a large solitary yellow flowering head. It is often found with the algal genus *Chara*.

Megalodonta beckii flowers from June to August.

Subtribe Pectinidae

Herbs (in Illinois) or shrubs; leaves basal or basal and cauline, usually opposite, pinnately lobed or divided; heads radiate (in Illinois) or discoid, with or without bractlets subtending the flowering heads; involucre hemispheric, campanulate, or turbinate; phyllaries in 1–2 series, free or connate, subequal; receptacle flat, concave, or convex, epaleate; ray flowers when present up to 20, variously colored, pistillate, fertile; disc flowers numerous, tubular, variously colored, bisexual, fertile, the corolla 5-lobed; cypselae glabrous or pubescent; pappus of up to 50 bristles or scales in 1–2 series.

Approximately 22 genera and 225 species are in this subtribe. Only 2 genera occur in Illinois.

1. Phyllaries free from each other; bractlets subtending the flowering heads 1 to few in number . . . . 59. *Dyssodia*
1. Phyllaries connate; bractlets absent . . . . 60. *Tagetes*

59. **Dyssodia** Cav.—Dogweed; Fetid Marigold

Annual (in Illinois) or perennial herbs; leaves opposite, pinnately divided, glandular, aromatic; heads radiate, several in corymbs, subtended by 1 to several bractlets; involucre campanulate or hemispheric; phyllaries in 1 series, free from each other; receptacle convex, epaleate; ray flowers 5–8, yellow, pistillate, fertile; disc flowers 50–100, tubular, yellow, bisexual, fertile, the corolla 5-lobed; cypselae obpyramidal, slightly angular; pappus of 15–20 scales in 2 series.

There are four species in this genus, all native to North America. All are strongly aromatic.

1. **Dyssodia papposa** (Vent.) A. Hitchc. Trans. Acad. Sci. St. Louis 5:503. 1891. Fig. 104.

Tagetes papposa Vent. Descr. Pl. Nuv. plate 36. 1801.

Boebera chrysanthemoides Willd. Sp. Pl. 3:2125. 1804.

Dyssodia chrysanthemoides (Willd.) Lag. Gen. & Sp. Nov. 29. 1816.

Boebera papposa (Vent.) Rydb. ex Britt. Man. 1012. 1901.

Annual herbs with fibrous roots; stems erect, branched, to 75 cm tall, glabrous or puberulent; leaves cauline, opposite, aromatic, pinnately divided, to 5 cm long, to 4 cm wide, the lobes serrate, glandular-dotted, glabrous or puberulent; heads radiate, several to numerous in corymbs, 6–10 mm across, subtended by 1 to several linear bractlets; involucre turbinate or campanulate; phyllaries in 1 series, free from each other, glandular-dotted; receptacle flat, epaleate; ray flowers 5–8, yellow,

1.5–2.5 mm long, pistillate, fertile; disc flowers 50–100, tubular, yellow, bisexual, fertile, the corolla 5-lobed, 2.5–3.0 mm long; cypselae obpyramidal, slightly angular, pubescent, 3.0–3.5 mm long; pappus of 15–20 scales in 2 series, 1–3 mm long.

Common Names: Dogweed; fetid marigold.
Habitat: Disturbed soil, particularly along interstate highways.

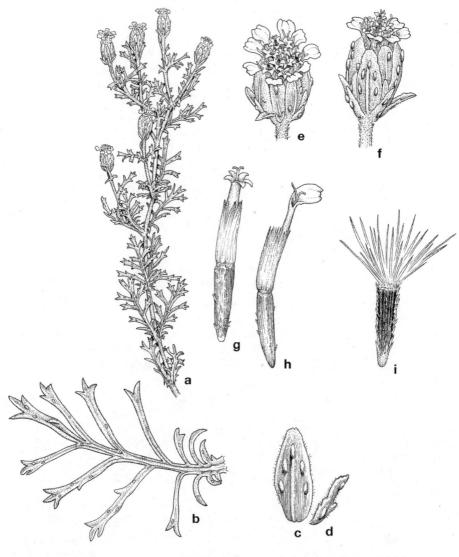

104. *Dyssodia papposa*
    (Dogweed).
a. Upper part of plant.
b. Leaf.

c. Outer phyllary.
d. Inner phyllary.
e, f. Flowering
    heads.

g. Disc flower.
h. Ray flower.
i. Cypsela.

Range: Native to the southern and western United States; adventive in Illinois.
Illinois Distribution: Scattered in Illinois, but increasing rapidly in the northeastern counties.

This species has a fetid odor. It resembles *Tagetes*, differing by its phyllaries that are free from each other and by the presence of bractlets subtending each flowering head.

Dyssodia papposa flowers during September and October.

60. **Tagetes** L.—Marigold

Annual (in Illinois) or perennial herbs; stems erect, branched; leaves cauline, opposite, usually pinnately divided; heads radiate (in Illinois) or discoid, 1 to several in corymbs, without subtending bractlets; involucre turbinate or campanulate; phyllaries in 1–2 series, connate at the base; receptacle convex to conical, epaleate; ray flowers when present up to 5–15 (in Illinois), rarely absent, yellow or orange, pistillate, fertile; disc flowers up to 100, tubular, usually yellow or orange, bisexual, fertile, the corolla 5-lobed; cypselae obpyramidal, glabrous or pubescent; pappus of 2–5 scales in 1 series, some of them erose or laciniate, some subulate to aristate.

About 40 species comprise this genus, native to tropical and warm temperate North and South America. Several are popular ornamentals, including double-flowered forms. Only one species has been found in the wild in Illinois.

1. **Tagetes erecta** L. Sp. Pl. 2:887. 1753. Fig. 105.
Annual herbs with fibrous roots; stems erect, branched, to 75 cm tall, glabrous; leaves cauline, opposite, pinnately divided into several leaflets, to 3 cm long, to 1 cm wide, glabrous or nearly so, segments lanceolate; head radiate, solitary, on peduncles up to 10 (–12) cm long, not subtended by bractlets; involucre campanulate; phyllaries in 1 or 2 series, connate at the base; receptacle convex, epaleate; ray flowers 8–15, yellow-orange, to 2 cm long, pistillate, fertile; disc flowers up to 100, tubular, yellow-orange, bisexual, fertile, the corolla 5-lobed, 7–12 mm long; cypselae obpyramidal, glabrous or strigillose, 6–10 mm long; pappus of 2 subulate scales up to 12 mm long and 2–4 linear scales 2–6 mm long.

Common Name: French marigold; African marigold.
Habitat: Disturbed soil.
Range: Native to Mexico; escaped from cultivation in several states.
Illinois Distribution: Known from Alexander and DuPage counties.

Although similar to *Dyssodia papposa*, *Tagetes erecta* differs by its connate phyllaries, the absence of bractlets subtending the flowering heads, and the much larger cypselae with 2 subulate scales and 2–4 linear scales.

This species is a common garden ornamental. Double-flowered variations are common in cultivated plants.

Tagetes erecta flowers from June to October.

105. *Tagetes erecta*
    (French marigold).
a. Upper part of plant.
b. Leaf.

c. Flowering head.
d. Ray flower.
e. Disc flower.

f. Flowering involucre.
g. Fruiting involucre.
h. Cypsela.

Subtribe Hymenopappinae Rydb. in Britt.

Annual, biennial, or perennial herbs; stems erect; leaves basal or basal and cauline, alternate, pinnately divided, glandular-dotted; heads discoid (in Illinois) or radiate, borne in corymbs, not subtended by bractlets; involucre hemispheric; phyllaries in 2 (–3) series, free from each other, subequal, often petaloid, with scarious margins; receptacle flat or convex, epaleate; ray flowers when present up to 8, or absent (in Illinois), pistillate, fertile; disc flowers up to 75, tubular, yellow, bisexual, fertile, the corolla 5-lobed; cypselae obconic to obpyramidal, 4-angled, conspicuously ribbed, glabrous or pubescent; pappus of up to 22 scales, or absent.

There are five genera in this subtribe. Only *Hymenopappus* occurs in the United States.

61. Hymenopappus L'Her.

Biennial or perennial herbs; stems erect, angular, glabrous or pubescent; leaves basal or basal and cauline, alternate, pinnately divided, glabrous or pubescent, glandular-dotted; heads discoid (in Illinois) or radiate, 1 to several in a corymb, not subtended by bractlets; involucre hemispheric; phyllaries in 2 (–3) series, 6–12 in number, subequal, appressed, petaloid, with scarious margins; receptacle convex, epaleate; ray flowers absent (in Illinois) or up to 8, yellow, pistillate, fertile; disc flowers up to 75, tubular, white (in Illinois) or yellow, bisexual, fertile, the co-rolla 5-lobed; cypselae obpyramidal, 4- or 5-angular, conspicuously 1- to 4-ribbed, glabrous or pubescent; pappus of up to 22 scales, or absent.

Twelve species comprise this genus, all in North America and Mexico. Only the following occurs in Illinois.

1. Hymenopappus scabiosaeus L'Her. Hymenopappus, plate 1. 1788. Fig. 106.

Rothia carolinensis Lam. Journ. Hist. Nat. 1:16. 1792.
Hymenopappus carolinensis (Lam.) Porter, Mem. Torrey Club 5:338. 1894.

Biennial herbs; stems erect, branched, angular, to 80 cm tall, glabrous, or woolly when young; leaves basal and cauline, alternate, to 25 cm long, to 2.5 cm wide, pinnately divided, petiolate, glabrous above, woolly below, the segments linear to oblong, the uppermost leaves less divided and usually sessile; heads discoid, 1 to several in corymbs, 10–16 mm across, on peduncles up to 5 cm long, not subtended by bractlets; involucre hemispheric; phyllaries in 2 (–3) series, 6–12 in number, oval to ovate to oblong, bright white, glabrous or puberulent; recep-tacle convex, epaleate; ray flowers absent; disc flowers up to 80, tubular, white, bisexual, fertile, the corolla 5-lobed, 3.2–5.5 mm long; cypselae obpyramidal, 4- or 5-angled, 1- to 4-ribbed, short-pubescent, 3–5 mm long; pappus of 14–18 scales up to 1 mm long.

Common Name: Old plainsman.
Habitat: Savannas, sand prairies.

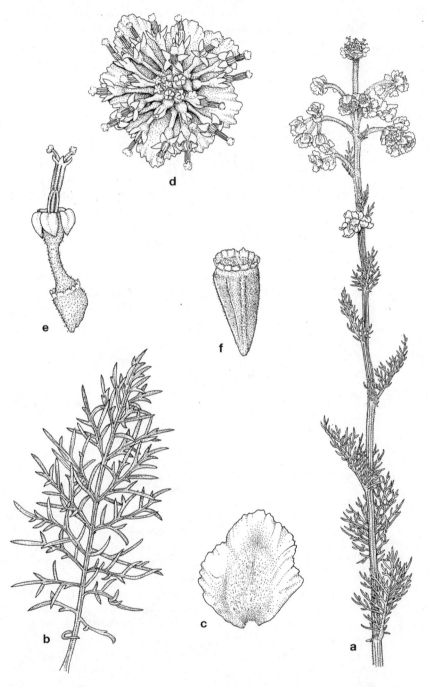

106. *Hymenopappus scabiosaeus* (Old plainsman).

a. Upper part of plant.
b. Leaf.
c. Phyllary.

d. Flowering head.
e. Disc flower.
f. Cypsela.

Range: Indiana to Nebraska, south to Texas and Florida.

Illinois Distribution: Very rare; known from Cass, Iroquois, Kankakee, and Mason counties.

This is the only species in the Asteraceae with bright white phyllaries and white disc flowers.

It is one of several rare species found in the sand prairies of Iroquois and Kankakee counties.

Hymenopappus scabiosaeus flowers from May to July.

Subtribe Gaillardiinae Less.

Annual or perennial herbs or shrubs; leaves basal or basal and cauline, alternate, unlobed or pinnately lobed, glandular-dotted; heads radiate, 1 to several in corymbs, not subtended by bractlets; involucre campanulate or globose or hemispheric; phyllaries in 2 (–3) series, equal or subequal; receptacle convex, flat, or conical, epaleate but sometimes with setae; ray flowers pistillate, fertile, or neutral, yellow, orange, or sometimes reddish; disc flowers few to numerous, bisexual, fertile, yellow, orange, or red-brown, the corolla 5-lobed; cypselae obpyramidal or obconic, usually pubescent; pappus of 2–12 scales.

There are twelve genera in this subtribe, all but one occurring in the United States. Three genera occur in Illinois.

1. Phyllaries reflexed in fruit; involucre globose or hemispheric.
    2. Stems unwinged; receptacle with setae .... 62. *Gaillardia*
    2. Stems usually winged; receptacle without setae.................... 63. *Helenium*
1. Phyllaries spreading to erect in fruit; involucre hemispheric or campanulate ....
    .... 64. *Tetraneuris*

62. **Gaillardia** Foug.—Blanket-flower

Annual or perennial herbs; stems erect, branched; leaves basal or basal and cauline, alternate, unlobed or pinnately lobed, pubescent, glandular-dotted; head radiate, solitary, not subtended by bractlets; involucre hemispheric; phyllaries in 2 (–3) series, reflexed in fruit, unequal to subequal, chartaceous; receptacle convex, epaleate but with setae; ray flowers up to 15, yellow, orange, red, or purple, pistillate, fertile; disc flowers up to 100, tubular, yellow, orange, purple, or red-brown, bisexual, fertile, the corolla 5-lobed; cypselae obpyramidal, angled, pubescent; pappus of 6–10 scales in 1–2 series.

There may be as many as twenty species in this genus, all native to the New World. Several species are grown as ornamentals. Three species have been found apparently in Illinois.

1. Ray and disc flowers yellow or purple (ray flowers sometimes red-purple near base); perennials.
    2. Flowering heads up to 4 cm across; stems puberulent; some of the leaves clasping at the base; cypselae 1.5–2.0 mm long .... 3. *G. aestivalis*
    2. Flowering heads at least 5 cm across; stems hirsute; none of the leaves clasping as the base; cypselae 2.5–6.0 mm long .... 1. *G. aristata*
1. Ray flowers yellow and red-purple; disc flowers purple; annuals .... 2. *G. pulchella*

I. **Gaillardia aristata** Pursh, Fl. Am. Sept. 2:573. 1813. Fig. 107.

Perennial herbs; stems erect, usually unbranched, up to 50 cm tall, densely hirsute; basal leaves simple, to 5 cm long, to 2.5 cm wide, oblong to elliptic, unlobed or pinnatifid, pubescent, on petioles up to 5 cm long; cauline leaves alternate, lanceolate to ovate-lanceolate, entire or dentate, rarely pinnatifid, pubescent, sessile or nearly so; head radiate, solitary, at least 5 cm across, on pubescent peduncles up to 25 cm long, not subtended by bractlets; involucre hemispheric; phyllaries in 2 (−3) series, subequal, ovate to lanceolate, acuminate at the apex, villous, to 15 mm long; receptacle convex, epaleate but with setae to 6 mm long; ray flowers up

107. *Gaillardia aristata*   a. Habit.   d. Ray flower.
     (Gaillardia).   b. Disc flower.   e, f. Cypselae.
                              c. Phyllary.

to 15, yellow, often with a red-purple base, to 15 (–18) mm long, 3-notched at the tip, pistillate, fertile; disc flowers up to 100, tubular, purple, bisexual, fertile, the corolla 5-lobed, 4–6 mm long; cypselae mostly obpyramidal, 2.5–6.0 mm long, villous near the base; pappus of 8 aristate scales 5–6 mm long.

Common Name: Gaillardia.
Habitat: Along a railroad.
Range: Native to the northwestern United States; occasionally adventive elsewhere.
Illinois Distribution: Jackson County: along a railroad, near Etherton.

This large-flowered species differs from the similar *G. aestivalis* by its perennial habit, its longer cypselae, and its densely hirsute stems.
   Gaillardia aristata flowers from June to August.

   2. **Gaillardia pulchella** Foug. Hist. Acad. Roy. Sci. Mem. Math. Phys. 1786:5. 1788. Fig. 108.
   Annual herbs with fibrous roots; stems erect to ascending, branched, to 50 cm tall, hirsute; basal leaves spatulate, unlobed or less commonly pinnatifid, hirsute, to 8 cm long, to 3 cm wide, petiolate, the middle and upper leaves pinnatifid or the uppermost unlobed and linear-lanceolate, sessile; head radiate, solitary, to 4.5 cm across, the peduncle to 15 cm long, without subtending bractlets; involucre hemispheric; phyllaries in 2 (–3) series, lanceolate, acuminate at the apex, usually hirsute, to 15 mm long; receptacle convex, epaleate, but with setae up to 3 mm long; ray flowers up to 20, red-purple and yellow, to 3 cm long, pistillate, fertile; disc flowers up to 100, tubular, purple, bisexual, fertile, the corolla 5-lobed, 3–5 mm long; cypselae obpyramidal, 2.0–2.5 mm long, villous; pappus of 8 aristate scales up to 7 mm long.

Common Name: Blanket-flower.
Habitat: Disturbed soil.
Range: Native to the central and southeastern United States.
Illinois Distribution: Escaped from cultivation and scattered in the state.

This handsome garden ornamental occasionally escapes from cultivation in Illinois. It is distinguished by its bicolored rays and its purple disc.
   Gaillardia pulchella flowers from June to September.

   3. **Gaillardia aestivalis** (Walt.) H. Rock, Rhodora 58:315. 1956. Fig. 109.
Helenium aestivalis Walt. Fl. Carol. 210. 1788.

Perennial herbs; stems erect, usually branched, to 60 cm tall, puberulent; leaves cauline, alternate, to 6 cm long, to 1.2 cm wide, linear-oblong to obovate, acute to acuminate at the apex, tapering to the sessile or short-petiolate base, or the uppermost clasping, sparsely serrate to entire, scabrellous; head radiate, solitary, up to 4 cm across, on a peduncle up to 15 cm long, not subtended by bractlets; involucre

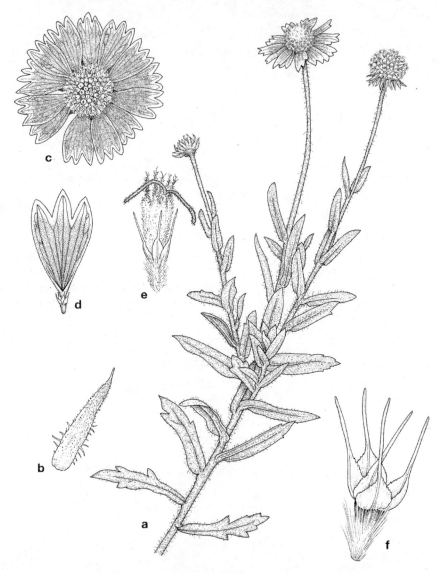

108. *Gaillardia pulchella*
(Blanket-flower).

a. Upper part of plant.
b. Phyllary.
c. Flowering head.

d. Ray flower.
e. Disc flower.
f. Cypsela.

hemispheric; phyllaries in 2 (–3) series, broadly lanceolate, to 15 mm long, scabrellous; receptacle convex, epaleate, but with setae up to 0.5 mm long, or absent; ray flowers up to 12, mostly yellow, 1–2 cm long, pistillate, fertile; disc flowers up to 75, tubular, yellow and/or purple, bisexual, fertile, the corolla 5-lobed, 4–7 mm long; cypselae obpyramidal, 1.5–2.0 mm long, angular, short-hairy at the base; pappus of up to 10 lanceolate scales 5–7 mm long.

109. *Gaillardia aestivalis*
(Gaillardia).

a. Upper part of plant.
b. Cluster of flowers.
c. Palea.

d. Ray flower.
e. Disc flower.
f. Cypsela.

Common Name: Gaillardia.
Habitat: Prairies.
Range: Illinois to Kansas, south to Texas and Florida.
Illinois Distribution: Apparently collected a single time in Alexander County.

Although there is a specimen labeled from Alexander County, Wunderlin specu-
lates that the label data may be in error. Until I can prove that this specimen was
not collected in Illinois, I am including it in the Illinois flora.

This species has the shortest cypselae of any *Gaillardia* in Illinois. It has mostly
all yellow flowering heads, whereas the other two species in Illinois have red-
brown color on the rays and disc.

Gaillardia aestivalis flowers from July to October.

63. **Helenium** L.—Sneezeweed

Annual or perennial herbs; stems erect, usually branched, usually winged; leaves
cauline, alternate, unlobed (in Illinois), glabrous or pubescent, glandular-dotted;
heads radiate, few to numerous in corymbs, pedunculate, not subtended by
bractlets; involucre globose; phyllaries in 2 (–3) series, reflexed in fruit, glandular-
dotted; receptacle conical to globose, epaleate, without setae; ray flowers up to 30,
yellow to red-brown, usually notched at the tip, pistillate, fertile or neutral; disc
flowers very numerous, tubular, yellow or purple, bisexual or fertile, the corolla
4- or 5-lobed cypselae obpyramidal to turbinate, 4- or 5-angled, conspicuously
ribbed, glabrous or pubescent; pappus of 5–8 (–12) acuminate or aristate scales.

There are thirty-two species in the genus, all native to the New World.

Helenium differs from *Gaillardia* by its usually winged stems and the receptacle
without setae. It differs from *Tetraneuris* by its winged stems, its globose involucre,
and its phyllaries that are reflexed in fruit.

Three species occur in Illinois.

1. Leaves linear to linear-filiform; stems not winged; plants annual . . . . 1. *H. amarum*
1. Leaves linear-lanceolate to ovate; stems winged; plants perennial.
    2. Disc globose, purplish . . . . .2. *H. flexuosum*
    2. Disc depressed-globose, yellow. . . . . 3. *H. autumnale*

1. **Helenium amarum** (Raf.) H. Rock, Rhodora 59:131. 1957. Fig. 110.
Gaillardia amara Raf. Fl. Ludov. 69–70. 1817.
Helenium tenuifolium Nutt. Journ. Phila. Acad. 7:66. 1834.

Annual herbs with a taproot; stems erect, branched, unwinged, to 70 cm tall,
glabrous or puberulent; leaves alternate, simple, linear to linear-filiform, to 3 cm
long, entire, unlobed (in Illinois), usually glabrous, sessile; heads radiate, several
to numerous, borne in corymbs, 1.5–2.5 cm across, on glabrous or puberulent peduncles up to 10 cm long, not subtended by bractlets; involucre globose; phyllaries
in 2 series, linear, pubescent, spreading to reflexed; receptacle globose, epaleate,
without setae; ray flowers up to 8, yellow, pistillate, fertile, to 15 mm long, usually
3-notched at the summit, spreading to slightly reflexed; disc flowers numerous,

110. *Helenium amarum* (Bitterweed).

a. Habit.
b. Cluster of phyllaries.
c. Flowering head.
d. Ray flower.
e. Disc flower.
f. Cypsela.

tubular, yellow or brownish yellow, bisexual, fertile, the corolla 5-lobed, 1.5–2.5 mm long; cypselae obpyramidal, 0.7–1.2 mm long, villous, conspicuously ribbed; pappus of 6–8 ovate, aristate scales.

Common Name: Bitterweed.
Habitat: Roadsides, old fields, other disturbed areas.
Range: Pennsylvania to Michigan and Kansas, south to Texas and Florida; California.
Illinois Distribution: Occasional in southern Illinois, extending north to Champaign and Pike counties where it may be adventive.

This species is readily distinguished by its short stature, wingless stems, and very narrow leaves.

Helenium amarum is fairly common along roadsides in the southern fourth of the state. It flowers from August to October.

    2. **Helenium flexuosum** Raf. New Fl. 4:81. 1838. Fig. 111.
Helenium nudiflorum Nutt. Trans. Am. Phil. Soc. (II) 7:384. 1841.
Leptopoda brachypoda Torr. & Gray, Fl. N. Am. 2:388. 1842.
Helenium polyphyllum Small, Fl. S.E.U.S. 1291. 1903.

Perennial herbs with fibrous roots; stems erect, branched, winged, to 1 m tall, puberulent to hirtellous to villous; leaves alternate, simple, to 3 cm long, to 1.5 cm wide, the lower spatulate, obtuse at the apex, entire to dentate, scabrous, on margined petioles, the middle and upper leaves linear-lanceolate to lanceolate, usually entire or sparsely serrate, scabrous, sessile; heads radiate, several to numerous, borne in corymbs, 2–3 cm across, on pubescent peduncles up to 10 cm long, not subtended by bractlets; involucre globose to ovoid; phyllaries in 2 series, connate at the base, pubescent, eventually reflexed; receptacle globose to conical to ovoid, epaleate, without setae; ray flowers up to 12, yellow, notched at the summit, to 2 cm long, pistillate, neutral; disc flowers numerous, tubular, purple, bisexual, fertile, the corolla 5-lobed, 2.5–4.0 mm long; cypselae obpyramidal, 1.0–1.2 mm long, pubescent; pappus of 5–6 lanceolate to ovate aristate scales up to 1.5 mm long.

Common Name: Purple-headed sneezeweed.
Habitat: Fields, sandy meadows, prairies, roadside ditches.
Range: Quebec to Wisconsin, south to Texas and Florida.
Illinois Distribution: Occasional in the southern half of Illinois, less common in northern Illinois, where it may be adventive.

This species is distinguished by its winged stems, yellow ray flowers, and purple disc flowers.

There is some variation exhibited in this species. Plants with very few cauline leaves have been called *H. nudiflorum*. Plants with numerous cauline leaves and a more ovoid receptacle have been called *H. polyphyllum*.

Helenium flexuosum flowers from June to September.

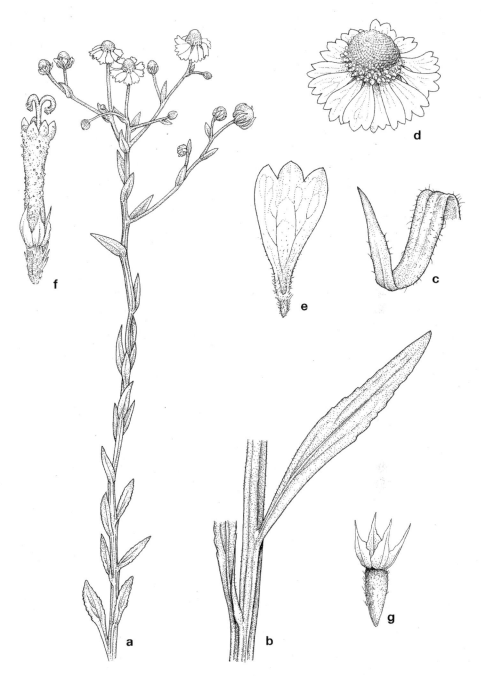

111. *Helenium flexuosum* (Purple-headed sneezeweed).

a. Upper part of plant.
b. Winged stem with leaves.
c. Phyllary.

d. Flowering head.
e. Ray flower.
f. Disc flower.
g. Cypsela.

3. **Helenium autumnale** L. Sp. Pl. 2:886. 1753. Figs. 112a, 112b, 112c.
Helenium pubescens Ait. Hort. Kew. 3:287. 1789.
Helenium altissimum Link ex Rydb. N. Am. Fl. 35:126. 1915.

Perennial herbs with fibrous roots; stems erect, usually branched, winged, to 1.5 m tall, glabrous or puberulent; leaves cauline, alternate, simple, unlobed, linear to elliptic to oblong, acute to acuminate at the apex, tapering to the sessile or short-petiolate base, membranous to stiff, to 25 cm long, to 5.5 cm wide, dentate or serrate or entire; heads radiate, 1 to many corymbs, on glabrous or pubescent peduncles up to 10 cm long, not subtended by bractlets; involucre globose; phyllaries in 2 series, connate below, pubescent; receptacle globose to ovoid, epaleate, without setae; ray flowers 8 or more, yellow, 3-notched at the summit, 3–25 mm long, 2–12 mm wide, pistillate, fertile, soon reflexed; disc depressed-globose, 8–23 mm across, with numerous tubular flowers, yellow, bisexual, fertile, the corolla 5-lobed, 2.5–4.0 mm long; cypselae turbinate, angular, pubescent, 1–2 mm long; pappus of 5–7 lanceolate, aristate scales 0.5–1.8 mm long.

Three varieties occur in Illinois:

a. Leaves up to 7 times longer than wide, usually more than 18 mm wide, membranaceous.
    b. Rays, or most of them, more than 13 mm long, 7–12 mm wide; disc 16–23 mm across . . . . 3a. *H. autumnale* var. *autumnale*
    b. Rays up to 13 mm long, 3–7 mm wide; disc 8–15 mm across . . . .
    . . . . 3b. *H. autumnale* var. *parviflorum*
a. Leaves more than 7 times longer than wide, not more than 18 mm wide, stiff . . . .
    . . . . 3c. *H. autumnale* var. *canaliculatum*

3a. **Helenium autumnale** L. var. **autumnale** Fig. 112a.
Leaves up to 7 times longer than wide, 5–25 cm long, 18–25 mm wide, dentate to serrate, membranaceous; rays 16–25 mm long, 7–12 mm wide; disc 16–23 mm across.

Common Name: Yellow-headed sneezeweed.
Habitat: Wet meadows, marshes, calcareous fens, wet ditches.
Range: Quebec to Minnesota, south to Missouri and Virginia.
Illinois Distribution: Common throughout the state.

This is the more common variety of *H. autumnale* in the state, where it occupies most kinds of wetland habitats. It is larger in all respects than the other two varieties.

This variety flowers from July to November.

3b. **Helenium autumnale** L. var. **parviflorum** (Nutt.) Fern. Rhodora 45:492. 1943. Fig. 112b.
Helenium parviflorum Nutt. Trans. Am. Phil. Soc., n.s. 7:384. 1841.

Leaves up to 7 times longer than wide, 7–35 mm long, 18–25 mm wide, often entire, membranaceous; rays 3–13 mm long, 3–7 mm wide; disc 8–15 mm across.

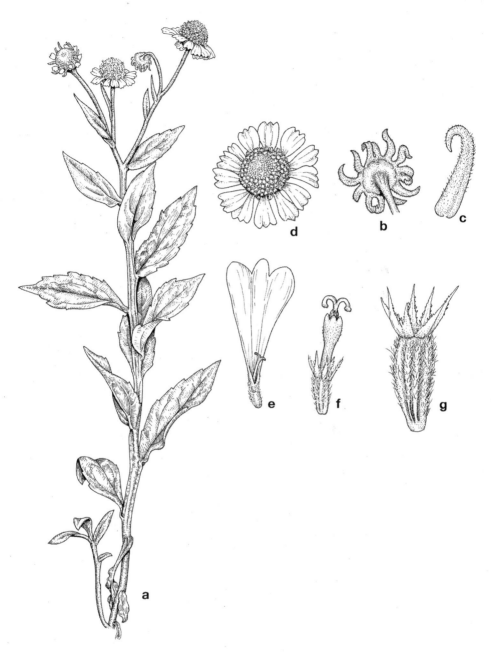

112a. *Helenium autumnale*
var. *autumnale*
(Yellow-headed
sneezeweed).

a. Upper part of plant.
b. Cluster of phyllaries.
c. Phyllary.
d. Flowering head.

e. Ray flower.
f. Disc flower.
g. Cypsela.

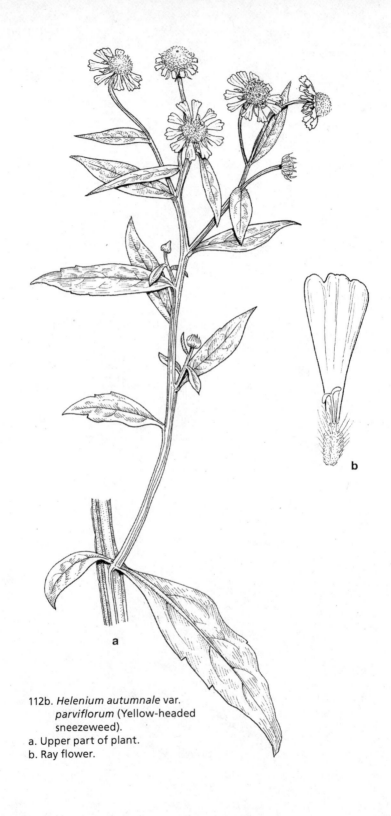

112b. *Helenium autumnale* var.
     parviflorum (Yellow-headed
     sneezeweed).
a. Upper part of plant.
b. Ray flower.

112c. *Helenium autumnale* var. *canaliculatum* (Yellow-headed sneezeweed).
a. Upper part of plant.
b. Flowering head.
c. Ray flower.
d. Disc flower.

Common Name: Small-flowered yellow-headed sneezeweed.
Habitat: Wet ground.
Range: Connecticut to Iowa, south to Illinois and Kentucky.
Illinois Distribution: Scattered in Illinois, but not common.

Nuttall described this plant as a new species, distinguishing it from *H. autumnale*. At first glance, it appears to be quite different from var. *autumnale* because of its narrow, entire leaves, its shorter and narrower rays, and its smaller disc.

This variety flowers from August to October.

3c. **Helenium autumnale** L. var. **canaliculatum** (Lam.) Torr. & Gray, Fl. N. Am. 2 (2):384. 1842. Fig. 112c.
Helenium canaliculatum Lam. Journ. Hist. Nat. 2:213. 1792.

Leaves more than 7 times longer than wide, not more than 18 mm wide, stiff.

Common Name: Narrow-leaved yellow-headed sneezeweed.
Habitat: Wet ground.
Range: Quebec to Minnesota, south to Nebraska, Missouri, and Ohio.
Illinois Distribution: Scattered in Illinois.

The stiff, rigid, very narrow leaves make this variety very different in appearance from var. *autumnale* and var. *parviflorum*. Lamarck described it as a different species.

This variety is scattered in Illinois wetlands. It flowers from August to October.

64. **Tetraneuris** Greene—Bitterweed; Starflower.

Annual or perennial herbs; stems erect (in Illinois) or decumbent, often unbranched; leaves all basal (in Illinois) or sometimes cauline and alternate, simple, often unlobed; heads radiate, borne singly or few in a corymb, the peduncles not subtended by bractlets; involucre hemispheric to campanulate; phyllaries in 3 series, free from each other, usually pubescent, appressed in fruit; receptacle hemispheric to conical, epaleate, without setae; ray flowers up to 25, yellow, 3-notched at the summit, usually 4-nerved, pistillate, fertile; disc flowers several to numerous, tubular, yellow, bisexual, fertile, the corolla 5-lobed; cypselae turbinate to obpyramidal, ribbed, pubescent; pappus of 4–8 acute to aristate scales.

About ten species are in the genus, all in North America and Mexico. Only the following occurs in Illinois.

1. **Tetraneuris herbacea** Greene, Pittonia 3:68. 1898. Fig. 113.
Actinella scaposa Nutt. var. *glabra* Gray, Man. Bot., ed. 5, 263. 1867.
Actinea herbacea (Greene) B.L. Robins. Rhodora 10:68. 1908.
Actinea acaulis (Pursh) Spreng. var. *glabra* (Gray) Cronq. Rhodora 47:403. 1948.
Hymenoxys acaulis (Pursh) K. F. Parker var. *glabra* (Gray) K.F. Parker, Madrono
    10:159. 1950.
Hymenoxys herbacea (Greene) Cronq. Man. Vasc. Pl. N.E.U.S., ed. 2, 1864. 1991.

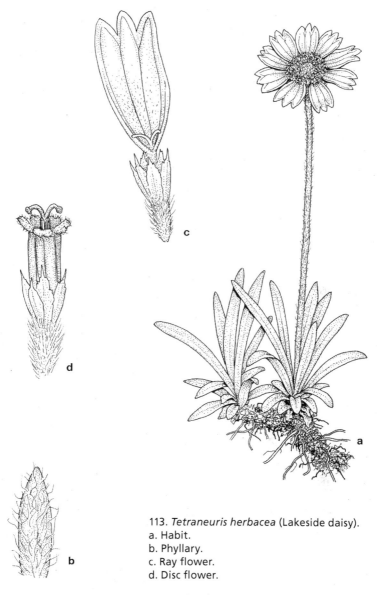

113. *Tetraneuris herbacea* (Lakeside daisy).
a. Habit.
b. Phyllary.
c. Ray flower.
d. Disc flower.

Perennial herbs with a thickened caudex; scape leafless, to 30 cm tall; leaves all basal, simple, crowded, linear to spatulate to oblanceolate, more or less rounded at the sessile base, 4–6 cm long, 6–10 mm wide, entire, 1-nerved, sparsely villous, densely punctate; head radiate, solitary, 3–4 cm in diameter, on a pubescent peduncle, not subtended by bractlets; involucre hemispheric; phyllaries in 3 series, unequal, obtuse at the apex, pubescent, somewhat scarious, up to 8 mm long; receptacle hemispheric, epaleate, without setae; ray flowers up to 25, yellow, up to 20 mm long, 3-notched at the summit, 1-nerved, pistillate, fertile; disc flowers up to 100, tubular, yellow, bisexual, fertile, the corolla 5-lobed, 3.5–4.0 mm long; cypselae turbinate or obpyramidal, 2.5–3.5 mm long, densely villous; pappus of usually 5 obovate scales to 2.2 mm long.

Common Name: Lakeside daisy; four-nerved starflower.
Habitat: Dolomite prairie.
Range: Illinois, Michigan, Ohio, Ontario.
Illinois Distribution: Very rare; known from Mason and Will counties.

This handsome species is distinguished by its large yellow flowering head on a leaf-less scape, with all the leaves crowded together at the base of the plant.

This is one of the rarer species in North America, known only naturally from two counties in Illinois, one in Michigan, one in Ohio, and a few locations in Ontario.

This species is sometimes placed in the genus *Hymenoxys*. It flowers during April and May.

Tribe Senecioneae Cass.

Annual or perennial herbs (in Illinois); leaves basal, cauline, or basal and cauline, usually alternate, unlobed or pinnatifid; heads discoid or radiate, borne in a variety of inflorescences, or sometimes solitary, with or without bractlets subtending the flowering heads; involucre cylindric to campanulate to turbinate; phyllaries in 1–2 series, rarely more, more or less equal, sometimes with scarious margins; receptacle flat or convex, epaleate; ray flowers absent or present and pistillate and fertile, usually yellow or orange; disc flowers numerous, tubular, bisexual, fertile, yellow or orange, the corolla 5-lobed; cypselae columnar, prismatic, or compressed, ribbed or nerved, glabrous or pubescent; pappus of 1–5 series of usually persistent smooth or barbellate bristles, sometimes of scales, rarely absent.

This tribe consists of approximately 120 genera and more than 3,200 species. Of these, 29 genera and nearly 170 species are in North America. Seven genera and 16 species are known from Illinois.

The characters of the Senecioneae include an involucre often subtended by bractlets, nearly equal phyllaries in usually 1–2 series, an epaleate receptacle, and a pappus of bristles or scales.

Key to the Genera of Tribe Senecioneae in Illinois
1. Ray flowers present, yellow or orange.
   2. Leaves simple, entire, serrate, or shallowly lobed.
      3. Head solitary . . . . 70. *Tussilago*
      3. Heads several . . . . 66. *Packera*
   2. Leaves simple and deeply divided or compound.
      4. Stems glandular-pubescent; rays 1–3 mm long . . . . 65. *Senecio*
      4. Stems eglandular; rays 6–10 mm long.
         5. Leaves once-pinnatifid . . . . 66. *Packera*
         5. Leaves 2- to 3-pinnatifid . . . . 65. *Senecio*

1. Ray flowers absent; flowers yellowish, pink, purple, or white.
  6. Leaves simple, entire, serrate, or shallowly lobed.
    7. Flowers yellowish. . . . . 65. *Senecio*
    7. Flowers pink, purple, or white.
      8. Flowers pink or purple . . . . 71. *Petasites*
      8. Flowers white.
        9. Phyllaries in 1 series; leaves not hastate.
          10. Calyx-like bracts at base of phyllaries; leaves basal. . . . . 71. *Petasites*
          10. Calyx-like bracts absent at base of phyllaries; leaves cauline.
            11. Phyllaries 5; flowers 5 per head. . . . . 68. *Arnoglossum*
            11. Phyllaries more than 5; flowers more than 5 per head. . . 67. *Erechtites*
        9. Phyllaries in 2 series; leaves hastate. . . . . 68. *Hasteola*
  6. Leaves deeply pinnatifid or pinnately compound . . . . 65. *Senecio*

65. **Senecio** L.—Groundsel

Annual or perennial herbs (in Illinois); stems branched or unbranched; leaves
cauline, alternate, toothed, lobed or pinnatifid or pinnately divided, glabrous or
pubescent; heads discoid or radiate, numerous, in corymbs, subtended by 2 to
several bractlets; involucre cylindric to campanulate to turbinate; phyllaries persistent, 5 to several in 1 or 2 series, free from each other, equal or nearly so, with a
scarious margin; receptacle flat to convex, epaleate; ray flowers 5 to several, pistillate, fertile, yellow, sometimes absent; disc flowers up to 100, tubular, bisexual, fertile, yellow, the corolla 5-lobed; cypselae cylindric or prismatic, 5-ribbed, glabrous
or pubescent; pappus persistent, of 30 or more smooth or barbellate bristles.

Approximately 1,000 species are in the genus, found worldwide. There are
three species in Illinois, all non-native. Native species, in the past included in *Senecio*, are now considered to be in a different genus known as *Packera*.

Senecio differs from *Packera* by usually having taproots or branched caudices
and by having inconspicuous or no ray flowers, although *S. jacobaea* is an excep-
tion to this, and by usually having black-tipped phyllaries, by having persistent
pappus, and by having all leaves cauline.

1. Rays absent . . . . 1. *S. vulgaris*
1. Rays present.
  2. Stems glandular-hairy; rays 1–2 mm long. . . . . 2. *S. viscosus*
  2. Stems glabrous or with cobwebby hairs; rays 6–12 mm long . . . . 3. *S. jacobaea*

1. **Senecio vulgaris** L. Sp. Pl. 2:867. 1753. Fig. 114.

Annual herbs with a taproot; stems ascending to erect, usually branched,
hollow, to 50 cm tall, glabrous or sparsely tomentose, at least when young; leaves
cauline, alternate, simple, mostly oblanceolate, to 10 cm long, to 3 cm wide,
dentate or pinnatifid, subacute to obtuse at the apex, tapering to the base, the
lowermost petiolate, the middle and upper leaves auriculate at the base, glabrous
or sparsely tomentose, at least when young; heads discoid, 8–20 in corymbs, to 6
mm across, subtended by up to 6 black-tipped bractlets 1.0–1.5 mm long; involucre cylindric, 8–10 mm high; phyllaries about 21, in 1–2 series, linear, sometimes

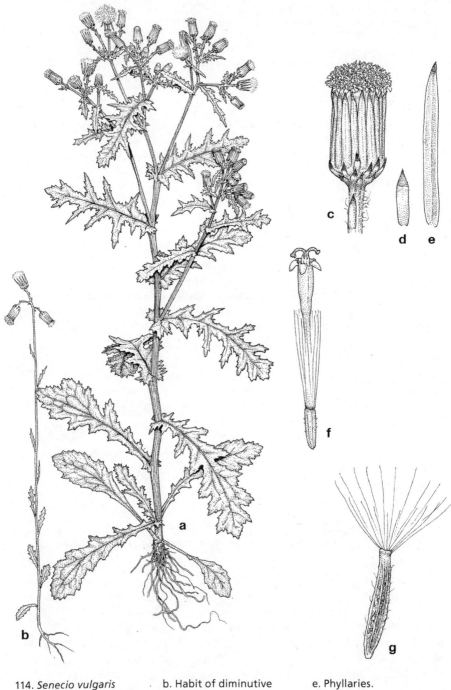

114. *Senecio vulgaris*
    (Common groundsel).
a. Habit.

b. Habit of diminutive
    plant.
c. Flowering head. d,

e. Phyllaries.
f. Disc flower.
g. Cypsela.

black-tipped, 4–6 mm long; receptacle flat to convex, epaleate; ray flowers absent; disc flowers up to 60, tubular, golden yellow, bisexual, fertile, the corolla 5-lobed; cypselae more or less columnar, several-ribbed, sparsely pubescent, 1.5–2.5 mm long; pappus persistent, of several white bristles.

Common Name: Common groundsel.
Habitat: Disturbed soil, often near nursery plots.
Range: Native to Europe and Asia; adventive in most of North America.
Illinois Distribution: Occasional in the northern half of Illinois.

This species is recognized by its black-tipped bractlets and phyllaries, the absence of ray flowers, and the auriculate base of the middle and upper leaves.
  Senecio vulgaris flowers from May to October.

  2. **Senecio viscosus** L. Sp. Pl. 2:868. 1753. Fig. 115.
  Annual ill-scented herbs with a taproot; stems ascending to erect, usually branched, not hollow, up to 50 cm tall, densely viscid pubescent with glandular hairs; leaves cauline, alternate, simple, oblong to spatulate, to 7.5 cm long, to 4 cm wide, pinnatifid, sometimes deeply so, tapering to the base, the lowermost peti-olate, the middle and upper leaves auriculate at the base, viscid pubescent, the hairs glandular; heads radiate, up to 20 (–30) in corymbs, to 6 mm across, on glandular-hirsute peduncles, subtended by up to 5 bractlets, the bractlets usually not black-tipped, 1–4 mm long; involucre cylindric, to 10 mm high; phyllaries up to 21, in 1–2 series, linear, usually black-tipped, 5–7 mm long; receptacle flat to convex, epaleate; ray flowers 12–20, yellow, recurved, 1–2 mm long, pistillate, fertile; disc flowers up to 60, tubular, yellow, bisexual, fertile, the corolla 5-lobed; cypselae more or less co-lumnar, several-ribbed, glabrous, 1.5–2.5 mm long; pappus of several white bristles.

Common Name: Fetid groundsel; sticky groundsel.
Habitat: Disturbed soil.
Range: Native to Europe and Asia; adventive in Canada and the northern United States.
Illinois Distribution: Known from Cook and Jackson counties. The Cook County specimen was collected in Bensenville in 1956.

This species is similar in appearance to *S. vulgaris*, but differs by its heavily viscid herbage and its very short and inconspicuous ray flowers.
  Senecio viscosus flowers from June to September.

  3. **Senecio jacobaea** L. Sp. Pl. 2:870. 175. Fig. 116.
  Jacobaea vulgaris Gaertn. Fruct. Sem. Pl. 2:445. 1791.

  Biennial herbs from branched caudices; stems erect, branched or unbranched, to 1.2 m tall, densely or sparsely cobwebby-hairy; leaves basal and cauline, the basal withering early, the cauline alternate, simple, to 25 cm long, to 7 cm wide,

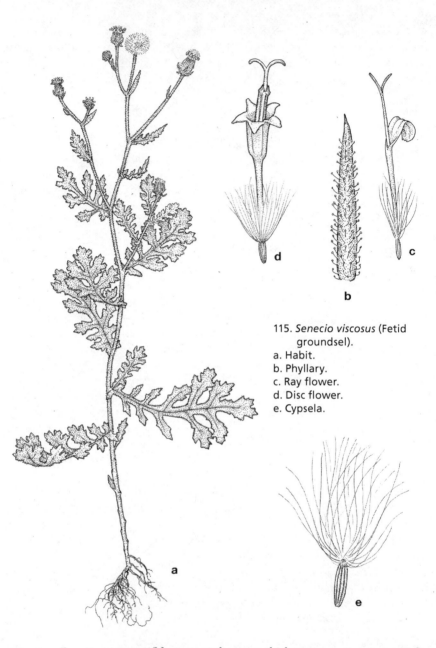

115. *Senecio viscosus* (Fetid groundsel).
a. Habit.
b. Phyllary.
c. Ray flower.
d. Disc flower.
e. Cypsela.

ovate, usually 2- to 3-pinnatifid, acute at the apex, the lowest tapering to a petiole, the uppermost sessile, cobwebby-hairy on the lower surface; heads radiate, up to 75, borne in corymbs, subtended by up to 6 bractlets 0.5–2.0 mm long; involucre campanulate to hemispheric; phyllaries 12–15, in 1–2 series, often black-tipped, 3.0–4.5 mm long, usually oblong, with a scarious margin; receptacle flat or convex, epaleate; ray flowers 12–15, yellow, 6–12 mm long, pistillate, fertile; disc

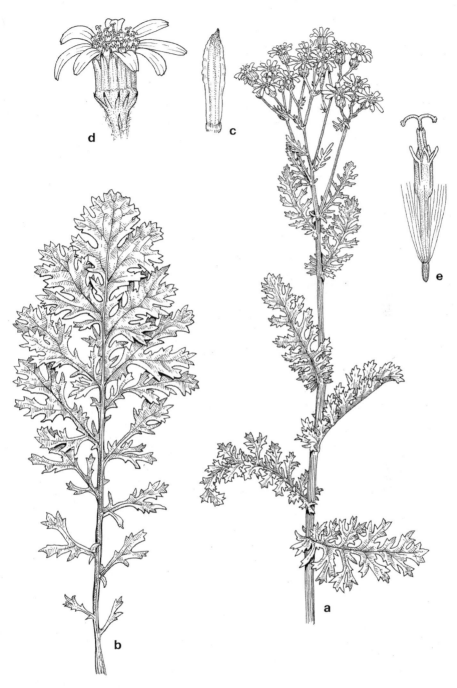

116. *Senecio jacobaea*
(Stinking Willie).

a. Upper part of
   plant.
b. Leaf.

c. Phyllary.
d. Flowering head.
e. Disc flower.

flowers up to 50, tubular, yellow, bisexual, fertile, the corolla 5-lobed; cypselae of the ray flowers usually glabrous, of the disc flowers pubescent, 1.5–3.0 mm long; pappus of several white bristles.

Common Name: Stinking Willie; tansy ragwort.
Habitat: Disturbed soil along a railroad.
Range: Native to Europe; adventive in a few of the United States.
Illinois Distribution: Only known from DuPage County where it was collected in
    1922 in Villa Park.

This non-native is the only species of *Senecio* in Illinois with conspicuous ray flowers. It has a foul odor when crushed. Wilhelm and Rericha (2017) call this species *Jacobaea vulgaris*.

    Senecio jacobaea flowers from July to September.

66. **Packera** A. Love & D. Love—Ragwort

Annual, biennial, or perennial herbs, with taproots or rhizomes; stems ascending to erect; leaves basal and/or cauline, alternate, the basal ones petiolate, the cauline ones often sessile, simple, unlobed or pinnately lobed or divided, the margins usually toothed; heads usually radiate (in Illinois), solitary or several in corymbs or cymes, usually subtended by bractlets; involucre cylindric to campanulate; phyllaries usually up to 21, 1–2 series, linear, not black-tipped, more or less equal, persistent, usually with scarious margins; receptacle flat, epaleate; ray flowers up to 13, yellow, pistillate, fertile; disc flowers up to 100, tubular, yellow, bisexual, fertile, the corolla 5-lobed; cypselae cylindric, conspicuously ribbed, glabrous or pubescent; pappus of several white, barbellulate bristles, caducous.

    There are about sixty-five species in this genus, almost all of them in North America. The species of *Packera* have been a part of *Senecio* in the past. The distinctions between *Packera* and *Senecio* are not often clear. Our species of *Packera* have distinct basal leaves (except *P. glabella*) and lack black-tipped phyllaries. Our species also have conspicuous ray flowers, but so does *Senecio jacobaea*.

    Six species are known from Illinois.

1. Leaves chiefly basal, the basal ones crenulate, dentate, or entire, the cauline ones (if present) sometimes pinnatifid.
  2. Basal leaves with winged petioles . . . . 3. *P. obovata*
  2. Basal leaves without winged petioles.
    3. Basal leaves cordate or subcordate or rounded at the base.
      4. Basal leaves deeply cordate at the base . . . . 1. *P. aurea*
      4. Basal leaves rounded or shallowly subcordate at the base . . . . . 6. *P. pseudaurea*
    3. Basal leaves tapering to the base, neither cordate nor subcordate.
      5. Leaves and stems floccose-tomentose at maturity, especially at the nodes; peduncles tomentose. . . . . 5. *P. plattensis*
      5. Leaves glabrous or nearly so; peduncles glabrous or nearly so. . . 4. *P. paupercula*
1. Leaves leafy to the summit; leaves all pinnatifid or coarsely sinuate-dentate . . . . . . 2. *P. glabella*

1. **Packera aurea** (L.) A. Love & D. Love, Bot. Not. 128:520. 1976. Fig. 117.
Senecio aureus L. Sp. Pl. 2:870. 1753.
Senecio gracilis Pursh, Fl. Am. Sept. 2:529–530. 1814.
Senecio aureus L. var. *gracilis* (Pursh) Britt. Ill. Fl. N.U.S. 3:481. 1898.
Senecio aureus L. var. *intercursus* Fern. Rhodora 45:499–500. 1943.

Perennial herbs, usually with rhizomes; stems erect, to 1 m tall, glabrous or sparingly tomentose in the axils of the cauline leaves; leaves basal and cauline, the basal ones to 15 cm long, to 6 cm wide, obtuse at the apex, deeply cordate at the base, crenate, usually glabrous, on petioles up to 20 cm long, the cauline lanceolate to oblong, to 6 cm long, usually pinnatifid, at least the uppermost sessile; heads radiate, up to 20 in a corymb, up to 2 cm across, the peduncles floccose-tomentose, subtended by up to 5 inconspicuous bractlets; involucre cylindric, 8–10 mm high; phyllaries up to 21, in 1–2 series, 6–8 mm long, linear, glabrous or pubescent; receptacle flat, epaleate; ray flowers 10–13, golden yellow, 7–10 mm long, pistillate, fertile; disc flowers up to 75, tubular, yellow, bisexual, fertile, the corolla 5-lobed, the lobes 2.0–2.5 mm long; cypselae cylindric, 8- to 10-ribbed, glabrous, 3.5–4.0 mm long; pappus of white barbellulate caducous bristles 4.5–5.5 mm long.

Common Name: Golden ragwort; squaw-weed.
Habitat: Wet ground, mesic woods, wet meadows, calcareous springy sites, fens.
Range: Labrador to Manitoba, south to Oklahoma, Mississippi, and Florida.
Illinois Distribution: Occasional throughout Illinois.

This species differs from other species of *Packera* in Illinois by its deeply cordate basal leaves. It is found in a variety of wetland habitats.

Short plants less than 50 cm tall and with the petioles of the basal leaves less than 10 cm long have been called var. *gracilis*. Plants over 50 cm tall with the petioles of the basal leaves more than 10 cm long have been called var. *intercursus*. Neither of these varieties has been transferred to *Packera*.

Packera aurea flowers from mid-April to mid-June.

2. **Packera glabella** (Poir. in Lam.) C. Jeffrey, Kew Bull. 47:101. 1992. Fig. 118.
Senecio lyratus Michx. Fl. Bor. Am. 2:120. 1803, *non* L. (1753).
Senecio glabellus Poir. in Lam. Encycl. 7:102. 1806.
Senecio lobatus Pursh, Syn. 2:436. 1807.

Annual or biennial herbs with fibrous roots; stems erect, soft, hollow, to 80 cm tall, strongly ribbed, glabrous except sometimes with cobwebby hairs in the leaf axils; cauline leaves alternate, to 15 cm long, decreasing in size toward the top of the stem, deeply pinnatifid, the segments oblong to obovate, crenate or sometimes entire, glabrous, occasionally subclasping at the base; heads radiate, up to 50, borne in corymbs, the peduncles sometimes with cobwebby hairs, subtended by

117. *Packera aurea* (Golden
    ragwort).
a. Habit.

b. Basal leaf.
c. Cluster of phyllaries.
d. Phyllary.

e. Flowering head.
f. Ray flower.
g. Disc flower.

several bractlets up to 4 mm long; involucre cylindric, 5–6 mm high; phyllaries up to 21, in 1–2 series, linear, glabrous, 5–7 mm long; receptacle flat, epaleate; ray flowers up to 13, yellow, 7–10 mm long, pistillate, fertile; disc flowers up to 60, tubular, yellow, bisexual, fertile, the corolla 5-lobed, the lobes 2–3 mm long; cypselae cylindric, ribbed, hispidulous or less commonly glabrous, 1.0–1.5 mm long; pappus of numerous white, barbellulate bristles 3–4 mm long.

118. *Packera glabella*
   (Butterweed).
a. Habit.

b. Basal rosette of
   leaves.
c. Cluster of phyllaries.
d. Phyllary.

e. Flowering head.
f. Ray flower.
g. Disc flower.
h. Cypsela.

Common Name: Butterweed.
Habitat: Wet fields, wet meadows, marshes, along streams, floodplain woods, roadsides.
Range: Ohio to Nebraska, south to Texas and Florida.
Illinois Distribution: Common in the southern half of Illinois; adventive in northern counties.

This showy species often forms many acres of bright yellow flowering heads in the spring in southern Illinois. It is readily recognized by its hollow, ribbed stems and its deeply pinnatifid leaves.

For many years, this species was known as *Senecio glabellus*.

Packera glabella flowers from April to early June.

3. **Packera obovata** (Muhl. ex Willd.) W.A. Weber & A. Love, Phytologia 49:47. 1981. Fig. 119.

Senecio obovatus Muhl. ex Willd. Sp. Pl. 3:1999. 1803.

Senecio aureus L. var. *obovatus* (Muhl. ex Willd.) Torr. & Gray, Fl. N. Am. 2:442. 1843.

Perennial herbs with very slender stolons and rhizomes; stems erect, often several, to 50 cm tall, glabrous or tomentose in the leaf axils; leaves mostly basal, forming a rosette, mostly obovate, sometimes suborbicular, to 10 cm long, to 8 cm wide, obtuse at the apex, tapering to the winged petiolate base, crenate or less commonly serrate, glabrous, sometimes purplish; cauline leaves few, lyrate-pinnatifid, glabrous, sessile; heads radiate, up to 15 in a corymb, 10–15 mm across, the peduncles glabrous or tomentose near the tip, subtended by several bractlets up to 4 mm long; involucre cylindric, 8–10 mm high; phyllaries up to 21, in 1–2 series, linear, glabrous or tomentose near the apex, 3–6 mm long; receptacle flat, epaleate; ray flowers often 12 but sometimes up to 21, yellow, 7–10 mm long, pistillate, fertile; disc flowers up to 70, tubular, yellow, bisexual, fertile, the corolla 5-lobed, the lobes 2–3 mm long; cypselae cylindric, ribbed, glabrous or hirsute, 1.0–1.5 mm long; pappus of numerous white, barbellulate bristles 3–6 mm long.

Common Name: Round-leaved groundsel; round-leaved ragwort.
Habitat: Rich woods, rocky outcrops, black oak savannas.
Range: Quebec to Ontario, then south across Michigan and Illinois to Kansas, then south to New Mexico and Florida.
Illinois Distribution: Scattered but rare in the southern three-fourths of the state; also Grundy County.

The distinguishing features of this species are its obovate or suborbicular basal leaves borne on winged petioles. None of the leaves is cordate.

Packera obovata flowers from April to June.

119. *Packera obovata*
(Round-leaved
groundsel).

a. Habit.
b. Basal rosette.
c. Cluster of phyllaries.
d. Palea.

e. Flowering head.
f. Ray flower.
g. Disc flower.
h. Cypsela.

4. **Packera paupercula** (Michx.) A. Love & D. Love, Bot. Not. 128:520. 1976.
Figs. 120a, 120b, 120c, 120d.
Senecio pauperculus Michx. Fl. Bor. Am. 2:120. 1803.

Perennial herbs with slender, branched rhizomes or stolons, or sometimes
without rhizomes and stolons but reproducing asexually by forming adventitious
rosettes from the roots; stems erect, up to 50 cm tall, glabrous or with tufts of
white hairs in the leaf axils; leaves basal and cauline, soft, the basal elliptic to lanceolate, to 10 cm long, to 2.5 cm wide, subacute to obtuse at the apex, tapering to
the wingless petiole, glabrous, pale green, rarely purplish, serrate to nearly entire,
the petiole usually longer than the blades, the cauline leaves alternate, lanceolate,
usually pinnatifid, acute at the apex, tapering to the sessile but non-clasping base,
glabrous or strongly tomentose, strongly reduced in size above; heads radiate, up
to 40, borne in more or less open corymbs, 10–18 mm across, on tomentose or
less commonly glabrous peduncles, subtended by a few bractlets 0.5–1.5 mm long;
involucre campanulate, up to 8 mm high; phyllaries about 13, sometimes as many
as 21, linear to linear-lanceolate, glabrous, up to 8 mm long; receptacle flat, epaleate; ray flowers up to 13, yellow, 9–10 mm long, pistillate, fertile; disc flowers up
to 75, tubular, yellow, bisexual, fertile, the corolla 5-lobed, the lobes 2–3 mm long;
cypselae cylindrical, ribbed, glabrous or less commonly hispidulous, 1–2 mm long;
pappus of numerous white, barbellulate bristles 3.5–4.5 mm long.

Four varieties, sometimes difficult to differentiate, occur in Illinois:
a. Rhizomes and stolons absent; plants reproducing asexually by adventitious rosettes
   arising from the roots . . . . 4d. *P. paupercula* var. *savannarum*
a. Rhizomes and stolons present; plants not reproducing asexually by adventitious
   rosettes arising from the roots.
   b. Plants with slender stolons . . . . 4c. *P. paupercula* var. *pseudotomentosa*
   b. Plants with short, thick rhizomes.
      c. Plants with up to 8 flowering heads and basal leaves up to 4 cm long . . . .
         . . . . 4a. *P. paupercula* var. *paupercula*
      c. Plants with more than 8 flowering heads and basal leaves 4 cm long or longer . . .
         . . . . 4b. *P. paupercula* var. *balsamitae*

4a. **Packera paupercula** (Michx.) A. Love & D. Love var. **paupercula** Fig. 120a.
Senecio paupercula Michx. Fl. Bor. Am. 2:120. 1803.

Plants with short, thick rhizomes, not reproducing asexually by adventitious
rosettes arising from the roots; flowering heads up to 8 per plant; basal leaves up
to 4 cm long.

Common Name: Northern ragwort.
Habitat: Wet prairies, moist sand flats, sedge meadows, open woods.
Range: Labrador to Alaska, south to Washington, Utah, Colorado, Illinois, and New
York.
Illinois Distribution: Common in the northern half of Illinois, rare in the southern
half.

120a. *Packera paupercula* var.
   paupercula (Northern
   ragwort).
a. Habit.
b. Node with leaf.

This variety differs from the similar-appearing *P. plattensis* by its leaves and stems floccose-tomentose at maturity and its tomentose peduncles.

    Packera paupercula var. *paupercula* flowers during May and June.

    4b. **Packera paupercula** (Michx.) A. Love & D. Love var. **balsamitae** (Muhl. ex Willd.) Mohlenbr. Phytoneuron 2015–67:1. 2015. Fig. 120b.
Senecio balsamitae Muhl. ex Willd. Sp. Pl. 1999. 1804.
Senecio aureus L. var. *balsamitae* (Muhl. ex Willd.) Torr. & Gray, Fl. N. Am. 2:442. 1843.
Senecio paupercula (Michx.) A. Love & D. Love var. *balsamitae* (Muhl. ex Willd.) Fern. Rhodora 23: 299. 1921.

Plants with thick, short rhizomes, not reproducing asexually by adventitious rosettes arising from the roots; flowering heads more than 8 per plant; basal leaves 4 cm long or longer.

Common Name: Balsam groundsel; balsam ragwort.
Habitat: Wet prairies, wet sand flats.
Range: Labrador to Alberta, south to Colorado, Alabama, and Georgia.
Illinois Distribution: Rare in the northern tier of counties.

The distinguishing characteristics of this variety sometimes overlap with var. *paupercula*.

    This variety flowers during May and June.

    4c. **Packera paupercula** (Michx.) A. Love & D. Love var. **pseudotomentosa** (Mack. & Bush) R. Kowal, Novon 18:224. 2008. Fig. 120c.
Senecio robbinsii Oakes ex Rusby var. *pseudotomentosa* Peck, Ann. Reg. Univ. State N.Y. 47:143. 1894.
Senecio pseudotomentosa (Peck) Mack. & Bush, Trans. Acad. Sci. St. Louis 12:88–89. 1902.
Plants with slender stolons, not reproducing asexually by adventitious rosettes arising from the roots.

Common Name: Northern ragwort.
Habitat: Rich woods, glades, old fields, sand barrens.
Range: Ontario to Wisconsin, south to Arkansas and Mississippi.
Illinois Distribution: Known from the northern half of the state.

This variety differs from other varieties of *P. paupercula* by its slender stolons, rather than thick, short rhizomes.

    Packera paupercula var. *pseudotomentosa* flowers during May and June.

    4d. **Packera paupercula** (Michx.) A. Love & D. Love var. **savannarum** R. Kowal, Novon 2:224. 2008. Fig. 120d.
Plants reproducing asexually by adventitious rosettes arising from the roots, without producing stolons or rhizomes.

120b. *Packera paupercula*
var. *balsamitae*
(Balsam ragwort).
a. Habit.

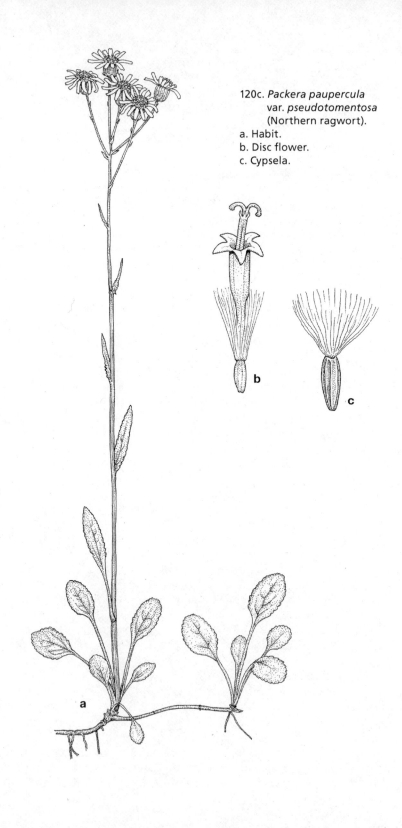

120c. *Packera paupercula*
var. *pseudotomentosa*
(Northern ragwort).
a. Habit.
b. Disc flower.
c. Cypsela.

120d. *Packera paupercula*
var. *savannarum*
(Savanna ragwort).

a. Habit.
b. Cluster of basal leaves.
c. Cluster of phyllaries.

d. Phyllary.
e. Ray flower.
f. Disc flower.

121. *Packera plattensis*
   (Prairie groundsel).
a. Habit.
b. Basal rosette of leaves.
c. Stem with base of
   cauline leaf.
d. Cluster of phyllaries.
e. Phyllary.
f. Flowering head.
g. Ray flower.
h. Disc flower.
i. Cypsela.

Common Name: Savanna ragwort.
Habitat: Bur oak savannas, mesic prairies, old fields, sand barrens.
Range: Illinois, Iowa, Michigan, Minnesota, Wisconsin.
Illinois Distribution: Known from Champaign, Grundy, Henry, and Winnebago
   counties.

This variety is quite different from the other varieties by lacking stolons and rhizomes, a feature also found in *P. plattensis*. Instead, it reproduces asexually by adventitious rosettes arising from roots.

This variety flowers during May and June.

   5. **Packera plattensis** (Nutt.) W.A. Weber & A. Love, Phytologia 49:48. 1981. Fig.
   121.
Senecio plattensis Nutt. Trans. Am. Phil. Soc., n.s. 7:413. 1841.

Plants reproducing asexually by adventitious rosettes arising from the roots, not forming stolons or rhizomes; stems several in a cluster, up to 50 cm tall, tomentose nearly throughout; leaves basal and cauline, firm, dark green, purple beneath, the basal oval to oblong to lanceolate, to 7.5 cm long, to 3 cm wide, usually pinnatifid, sometimes only dentate, acute at the apex, tapering or rounded at the petiolate base, usually floccose-tomentose, the cauline leaves alternate, pinnatifid, sparsely pubescent, the uppermost sessile; heads radiate, up to 40, in dense corymbs, 14–20 mm across, on tomentose peduncles, subtended by a few bractlets 0.5–2.0 mm long; involucre campanulate, 4–8 mm high; phyllaries 13 or 21, linear to linear-lanceolate, densely tomentose, 5–6 mm long; receptacle flat, epaleate; ray flowers 8–12, yellow, 9–10 mm long, pistillate, fertile; disc flowers up to 80, tubular, yellow, bisexual, fertile, the corolla 5-lobed, 3.5–4.5 mm long; cypselae cylindrical, ribbed, hirtellous or less commonly glabrous, 1.5–2.5 mm long; pappus of numerous white, barbellulate bristles 6.5–7.5 mm long.

Common Name: Prairie groundsel; prairie ragwort.
Habitat: Dry prairies, savannas, woods, fields, particularly in sandy areas.
Range: Ontario to Saskatchewan, south to New Mexico, Oklahoma, Louisiana, and
   Georgia.
Illinois Distribution: Common in northern and western Illinois, rare or absent
   elsewhere.

This species is similar to *P. paupercula*, and some specimens seem to be intermediate between the two. Both *P. plattensis* and *P. paupercula* var. *savannarum* reproduce asexually by adventitious rosettes that arise from the roots. The other three varieties of *P. paupercula* have stolons or rhizomes.

The following chart distinguishes *P. plattensis* from *P. paupercula*.

	Packera plattensis	*Packera paupercula*
underground parts	no stolons or rhizomes	stolons or rhizomes
stems	tomentose	mostly glabrous
basal leaves	floccose-tomentose	somewhat pubescent
all leaves	firm, dark green-purple beneath, acute at apex	soft, pale green, rarely purplish, subacute to obtuse at apex
corymb	densely crowded	more or less open
peduncles	tomentose	glabrous
phyllaries long	tomentose, 5–6 mm long	glabrous, up to 8 mm
ray flowers	8–10 in number	up to 13 in number
corolla lobe length	3.5–4.5 mm	2–3 mm
pappus length	6.5–7.5 mm	3.5–4.5 mm
habitat	usually dry	usually moist or wet

Packera plattensis occurs in sandy areas of prairies, woods, and fields. It flowers during May and June.

6. **Packera pseudaurea** (Rydb.) W.A. Weber & A. Love, Phytologia 49:48. 1981. var. **semicordata** (Mack. & Bush) Trock & T.M. Barkley, Sida 18:386. 1998. Fig. 122.

Senecio semicordatus Mack. & Bush, Ann. Mo. Bot. Gard. 16:107. 1905.
Senecio pseudaureus Rydb. var. *semicordatus* (Mack. & Bush) T.M. Barkley, Trans. Kansas Acad. Sci. 65: 341. 1962.

Perennial herbs with fibrous roots; stems erect, branched or unbranched, to 40 cm tall, glabrous or nearly so; leaves basal and cauline, the basal ones to 6 cm long, to 4 cm wide, broadly lanceolate to ovate, subacute to obtuse at the apex, rounded or barely subcordate at the base, crenate, glabrous or nearly so, shallowly pinnatifid to nearly entire, the petioles nearly twice as long as the blades, the cauline leaves lanceolate to oblong, to 4.5 cm long, becoming smaller toward the top of the stem, pinnatifid, at least the uppermost sessile or sometimes clasping; heads radiate, up to 12, in crowded corymbs, on glabrous peduncles, subtended by up to 5 inconspicuous bractlets; involucre cylindric, 8–10 mm high; phyllaries up to 25, in 1–2 series, linear, glabrous, 4–6 mm long; receptacle flat, epaleate; ray flowers up to 13 in number, yellow, 6–10 mm long, pistillate, fertile; disc flowers up to 10, tubular, yellow, bisexual, fertile, the corolla 5-lobed, the lobes 2–3 mm long; cypselae cylindric, ribbed, glabrous, 1.0–1.5 mm long; pappus of numerous white, barbellulate bristles 4.5–5.5 mm long.

Common Name: Western heartleaf groundsel.
Habitat: Wet prairies, fens, flatwoods.

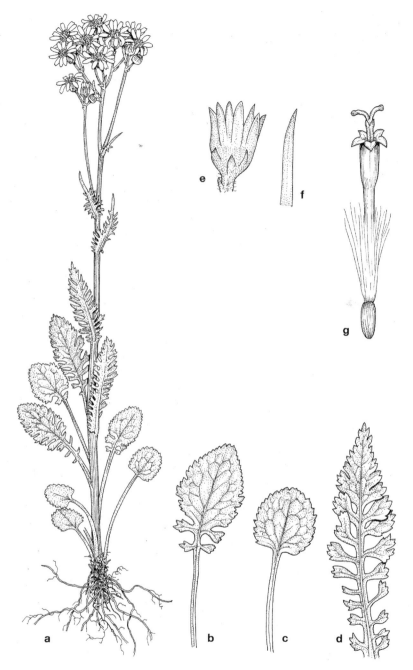

122. *Packera pseudaurea*
(Western heartleaf
groundsel).

a. Habit.
b. Leaf midway on
   stem.
c. Basal leaf.

d. Upper leaf.
e. Cluster of phyllaries.
f. Phyllary.
g. Disc flower.

Range: Minnesota to North Dakota, south to Kansas, Missouri, and Illinois.
Illinois Distribution: Known only from Cook and Lake counties.

This plant is closely related to *P. aurea* and for years it was considered to be a variety of *P. aurea*. The most obvious difference is that the basal leaves of *P. aurea* are deeply cordate, while the basal leaves of *P. pseudaurea* var. *semicordata* are rounded or barely subcordate. Typical var. *pseudaurea*, which is found in the far western United States, differs by having its basal leaves sharply dentate and its phyllaries 7–8 mm long.

Other differences between *P. aurea* and *P. pseudaurea* var. *semicordata* are summarized below.

	Packera aurea	*Packera pseudaurea* var. *semicordata*
underground	rhizomes	fibrous roots
basal leaves	deeply cordate	rounded at base or barely subcordate
flowering heads	up to 20	up to 12
peduncles	floccose-tomentose	glabrous to sparsely tomentose
phyllaries	6–8 mm long	4–6 mm long
rays	10–13 mm long	6–10 mm long
cypselae	3.5–4.0 mm long	1.0–1.5 mm long

Packera pseudaurea var. *semicordata* flowers in April and May.

67. **Erechtites** Raf.—Fireweed

Annual or perennial herbs, sometimes strongly aromatic, usually with a taproot; stems erect; leaves basal and cauline, the cauline alternate, pinnatifid or coarsely toothed; heads disciform, borne in corymbs or cymes, subtended by several bractlets; involucre cylindric to urceolate; phyllaries up to 21, usually in 1 series, equal, with scarious margins; receptacle flat or convex, epaleate; ray flowers absent; outer disc flowers narrowly tubular, up to 100 in 2 series, white to pale yellow, pistillate, fertile, the corolla 4- or 5-lobed; disc flowers tubular, up to 50, white to pale yellow, bisexual, fertile, the corolla 4- or 5-lobed; cypselae 5-angled or 5-ribbed, several nerved, glabrous or pubescent; pappus of numerous white, barbellulate bristles.

Erechtites consists of twelve species known from North America, South America, and the West Indies; New Zealand; Australia.

This genus differs from other genera in tribe Senecioneae by its whitish or pale yellow disc or disciform flowering heads and its 21 phyllaries arranged in a single series.

Only the following species occurs in Illinois.

1. **Erechtites hieracifolius** (L.) Raf. in DC. Prodr. 6:294. 1838. Figs. 123a, 123b, 123c.
Senecio hieracifolius L. Sp. Pl. 2:866. 1753.

Annual herbs with a taproot; stems erect, sometimes branched, grooved, to 3 m tall, glabrous or irregularly pubescent, semisucculent; leaves basal and cauline, all similar, or the uppermost very much smaller, very thin, ovate-lanceolate to oblong, irregularly toothed, sometimes deeply so, acute at the apex, tapering to the base, with a broad, sessile or nearly clasping base or tapering to the base, to 20 cm long, to 8 cm wide; heads disciform or discoid, up to 40, borne in corymbs, subtended by up to 10 bractlets; involucre urceolate, 10–15 mm high; phyllaries about 21, in 1 series, equal, linear, with scarious margins; receptacle flat to convex, epaleate; ray flowers absent; outer disc flowers up to 100, tubular, in 2 series, white to pale yellow, pistillate, fertile, the corolla 4- to 5-lobed; inner disc flowers up to 50, tubular, bisexual, fertile, white to pale yellow, the corolla 4- or 5-lobed; cypselae somewhat angular, with 10–12 ribs, strigose, 2–3 mm long; pappus of numerous bright white, barbellulate bristles 6–10 mm long.

Recently a few botanists have transferred this species to the genus *Erigeron*.

Three varieties have been found in Illinois:

a. Upper leaves with broad-based sessile or sometimes clasping base.
    b. All cauline leaves similar in size. . . . . 1a. *E. hieracifolius* var. *hieracifolius*
    b. Upper cauline leaves rapidly reduced in size . . . . . 1b. *E. hieracifolius* var. *intermedius*
a. Upper leaves tapering to base, sometimes short-petiolate . . . .
    . . . . 1c. *E. hieracifolius* var. *praealtus*

1a. **Erechtites hieracifolius** (L.) Raf. ex DC. var. **hieracifolius** Fig. 123a.
Upper leaves with broad-based sessile or sometimes clasping base; all cauline leaves similar in size.

Common Name: Fireweed.
Habitat: Usually disturbed woods and fields, marshes, sedge meadows.
Range: Native from Quebec to Minnesota, south to Texas and Florida; adventive elsewhere.
Illinois Distribution: Common throughout the state.

When fruiting, this species is easily recognized by its bright white pappus, particularly if the sun is shining on the plants.

The typical variety is common in disturbed areas throughout the state. It is one of the first plants to come into an area after a fire. It flowers from July to October.

1b. **Erechtites hieracifolius** (L.) Raf. in DC. var. **intermedius** Fern. Rhodora 19:27. 1917. Fig. 123b.
Upper leaves with broad-based sessile or sometimes clasping base; upper cauline leaves much smaller in size than the lower cauline leaves.

Common Name: Fireweed.
Habitat: Disturbed fields, marshes, sedge meadows.
Range: Prince Edward Island to Ontario, south to Texas and Florida.
Illinois Distribution: Scattered in Illinois.

123a. *Erechtites hieracifolius* var.
    hieracifolius (Fireweed).
a. Habit.
b. Leaf. var. *praealtus* (Fireweed).
c. Leaf.

123b. *Erechtites hieracifolius*
    var. *intermedius*
    (Fireweed).

d. Upper part of plant.
e. Middle part of plant.
f. Leaf

The rapidly reduced upper leaves are strikingly different in appearance from the typical variety. Both varieties may sometimes be found together.

This variety flowers from June to October.

1c. **Erechtites hieracifolius** (L.) Raf. ex DC. var. **praealtus** (Raf.) Fern. Rhodora 19:27. 1917. Fig. 123c.

Erechtites praealtus Raf. Fl. Ludov. 65. 1817.

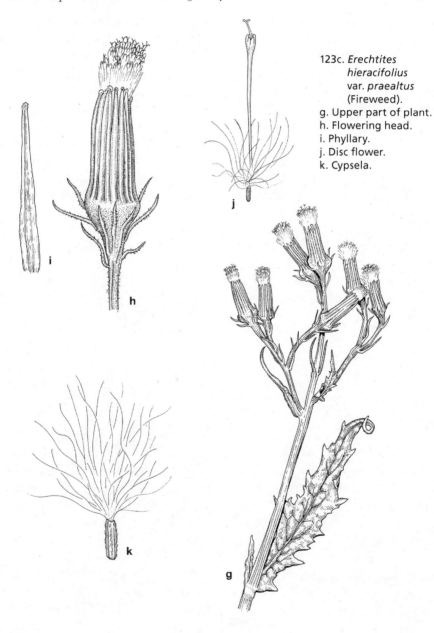

123c. *Erechtites hieracifolius* var. *praealtus* (Fireweed).
g. Upper part of plant.
h. Flowering head.
i. Phyllary.
j. Disc flower.
k. Cypsela.

Upper leaves tapering to the base, sometimes petiolate.

Common Name: Fireweed.
Habitat: Disturbed fields.
Range: Quebec to Wisconsin, south to Illinois, Tennessee, and Florida.
Illinois Distribution: Scattered but local, mostly in the southern third of Illinois.

Vegetatively this variety looks very different from either var. *hieracifolius* or var. *intermedius* because of its tapering and often petiolate leaves. In fact, Rafinesque described it as a new species. There are few floral differences, however. Most of the plants of this variety that I have seen are exceptionally tall, often reaching a height of three meters.

This variety flowers from July to October.

68. **Hasteola** Raf.—Hastate-leaved Indian Plantain

Perennial herbs with rhizomes; stems erect; leaves basal and cauline, the cauline alternate, usually hastate, doubly serrate; heads discoid, borne in corymbs, white or purplish, subtended by up to 10 bractlets; involucre cylindrical; phyllaries up to 15, in 2 series, oblong to linear, more or less equal; receptacle flat, epaleate; ray flowers absent; disc flowers up to 55, tubular, bisexual, fertile, the corolla 5-lobed; cypselae narrowly cylindric, several-ribbed, glabrous; pappus of numerous persistent, white, barbellulate bristles.

Two species comprise this genus, both in the United States. The only species that occurs in Illinois was at one time placed in *Cacalia* or *Synosma*.

Hasteola is similar to *Arnoglossum*, differing by having about 15 phyllaries per flowering head, rather than 5, and by having the phyllaries in 2 series, rather than 1. The flowering heads are subtended by bractlets in *Hasteola* while there are no bractlets in *Arnoglossum*.

Only the following species occurs in Illinois.

1. **Hasteola suaveolens** (L.) Pojarkova, Bot. Mater. Gerb. Bot. Inst. Komarova Akad. Nauk. S.S.S.R. 20:381. 1960. Fig. 124.
Cacalia suaveolens L. Sp. Pl. 2:835. 1753.
Synosma suaveolens (L.) Raf. ex Britt. Ill. Fl. N. U. S. 3:475. 1898.

Perennial herbs with rhizomes; stems erect, unbranched, to 1.5 m tall, glabrous or nearly so, often glaucous; leaves basal and cauline, reniform, triangular-lanceolate, often hastate, to 40 cm long, to 30 cm wide, acuminate at the apex, truncate or slightly tapering to the base, glabrous, sharply serrate, the lowermost sometimes on slightly winged petioles, the uppermost usually sessile; heads discoid, 3–5 mm across, borne in a corymb, subtended by up to 10 subulate bractlets; involucre campanulate, 10–15 mm high; phyllaries about 15, in 2 series, broadly linear, acute at the apex; receptacle flat, epaleate; ray flowers absent; disc flowers up to 55, tubular, pinkish, bisexual, fertile, the corolla 5-lobed; cypselae cylindrical, pale brown, glabrous, 5–8 mm long; pappus of numerous persistent, white, barbellulate bristles 6–7 mm long.

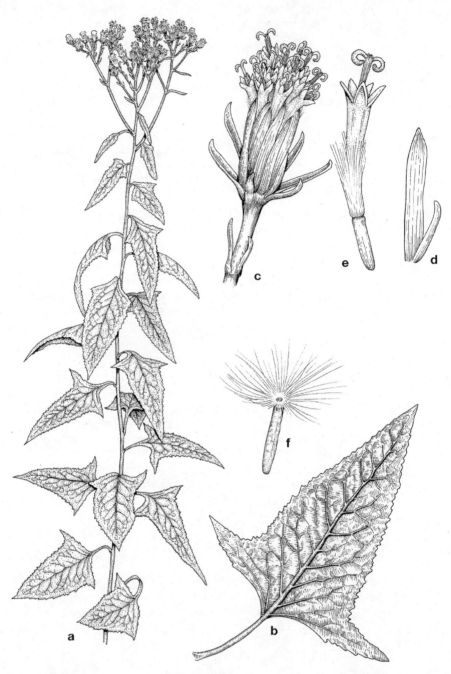

124. *Hasteola*
    suaveolens (Sweet-
    scented Indian
    plantain).

a. Upper part of plant.
b. Leaf.
c. Involucre with
   flowering heads.

d. Palea.
e. Disc flower.
f. Cypsela.

Common Name: Sweet-scented Indian plantain; hastate-leaved Indian plantain.
Habitat: Wet ground, calcareous fens.
Range: Massachusetts to Minnesota, south to Missouri, Tennessee, and North Carolina.
Illinois Distribution: Not common in northern Illinois, rare in southern Illinois.

Until 1960, this species had been called *Cacalia suaveolens* or, even earlier, *Synosma suaveolens*. It differs from the other species of *Cacalia* (now *Arnoglossum*) in Illinois by its 15 phyllaries arranged in 2 series, its hastate leaves, and the presence of bractlets at the base of the flowering heads.

Hasteola suaveolens flowers from July to September.

69. **Arnoglossum** Raf.—Indian Plantain

Perennial herbs with rhizomes or sometimes with a taproot; stems erect, often unbranched; leaves basal and cauline, alternate, simple, entire or occasionally toothed or lobed, glabrous, petiolate or sessile; heads discoid, borne in corymbs, not subtended by bractlets; involucre more or less cylindrical; phyllaries 5, in 1 series, equal, the margins scarious, the midvein sometimes winged; receptacle flat, epaleate, often with a central cusp to 2 mm long; ray flowers absent; disc flowers 5, tubular, usually white or cream, bisexual, fertile, the corolla 5-lobed; cypselae ellipsoid to fusiform, several-ribbed, glabrous; pappus of numerous white, usually scabrous bristles.

Arnoglossum consists of eight species, all in eastern North America. For many years, our plants were included in the genus *Cacalia*. Although Rafinesque named *Arnoglossum* in 1817, he later renamed it *Mesadenia*. Rafinesque described *Arnoglossum plantagineum* in 1817. Later, in 1836, he described *A. reniforme* and *A. atriplicifolium*, but in a genus he called *Mesadenia*. His reasoning is fairly valid since his species of *Arnoglossum* had firm, nearly entire, green leaves with parallel veins and with the midvein of the phyllaries winged, while his species of *Mesadenia* had membranaceous, coarsely dentate and often glaucous leaves with palmate venation, and with the midvein of the phyllaries unwinged.

Three species of *Arnoglossum* occur in Illinois:

1. Lower leaves reniform to deltate-ovate, as wide as long, lobed or coarsely angular-dentate, membranaceous, with palmate venation; midvein of phyllaries unwinged.
    2. Leaves glaucous on the lower surface; lower leaves deltate-ovate; stems not conspicuously grooved . . . . .1. *A. atriplicifolium*
    2. Leaves green on the lower surface; lower leaves reniform; stems conspicuously grooved . . . . 2. *A. reniforme*
1. Lower leaves lance-ovate, longer than wide, entire or shallowly dentate or crenate, firm, with parallel venation; midvein of phyllaries winged . . . . 3. *A. plantagineum*

1. **Arnoglossum atriplicifolium** (L.) H. Robins. Phytologia 28:294. 1974. Fig. 125.

Cacalia atriplicifolia L. Sp. Pl. 2:835. 1753.
Senecio atriplicifolius (L.) Hook. Fl. Bor. Am. 1:332. 1833.
Mesadenia atriplicifolia (L.) Raf. New Fl. 4:79. 1836.

Perennial herbs with rhizomes; stems erect, unbranched, to 2 m tall, glabrous, not conspicuously grooved; leaves membranaceous, basal and cauline, the basal deltate-ovate, acute at the apex, truncate or subcordate at the base, to 18 (−20) cm long, nearly as wide, glaucous on the lower surface, glabrous, palmately nerved, angular-lobed, long-petiolate, the cauline leaves rhombic to reniform, smaller than the basal, coarsely dentate, glaucous on the lower surface, glabrous, on shorter petioles or sessile; heads discoid, numerous, borne in corymbs, not subtended by

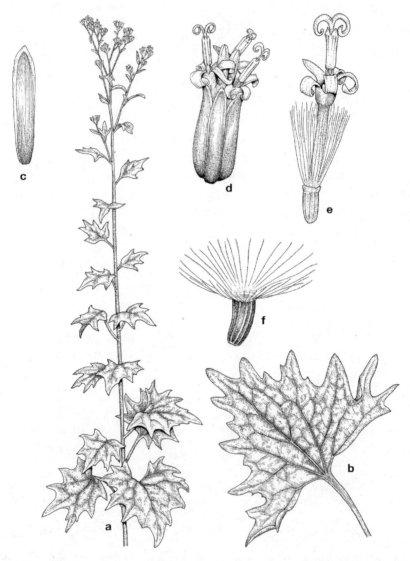

125. *Arnoglossum*
    atriplicifolium (Pale
    Indian plantain).

a. Upper part of plant.
b. Leaf.
c. Phyllary.

d. Involucre of flowers.
e. Disc flower.
f. Cypsela.

bractlets; involucre cylindric, 8–10 mm high; phyllaries 5, in 1 series, oblong, the midvein unwinged, with a scarious margin; receptacle flat, epaleate; ray flowers absent; disc flowers 5, tubular, white, bisexual, fertile, the corolla 5-lobed, up to 10 (–12) mm long; cypselae ellipsoid, several-ribbed, brown or purplish, glabrous, 4–5 mm long; pappus of numerous white, scabrous bristles 5–6 mm long.

Common Name: Pale Indian plantain.
Habitat: Moist woods, dry woods, prairies, dune slopes, fens, black oak savannas.
Range: New York to Minnesota, south to Nebraska, Oklahoma, Mississippi, and
   Florida.
Illinois Distribution: Occasional throughout the state.

Arnoglossum atriplicifolium is readily distinguished by its strongly glaucous lower leaf surface and its coarsely angular-dentate basal leaves. In the past, this species has usually been called *Cacalia atriplicifolia*.

   This species flowers from June to October and is found in a variety of habitats.

   2. **Arnoglossum reniforme** (Hook.) H. Robins. Phytologia 46:441. 1980. Fig.
   126.
Cacalia reniformis Muhl. ex Willd. Sp. Pl. 3:1735. 1803.
Senecio atriplicifolius (L.) Hook. var. *reniformis* Hook. Fl. Bor. Am. 1:332. 1834.
Mesadenia reniformis Raf. New Fl. 4:79. 1836.
Senecio muehlenbergii Sch.-Bip. Flora 28:499. 1845.
Mesadenia muehlenbergii (Sch.-Bip.) Rydb. Fl. Plains N. Am. 873. 1932.
Cacalia muehlenbergii (Sch.-Bip.) Fern. Rhodora 40:356. 1938.
Arnoglossum muehlenbergii (Sch.-Bip.) H. Robins. Phytologia 28:299. 1974.

   Perennial herbs with rhizomes; stems erect, unbranched, to 3 m tall, glabrous or nearly so, angled, conspicuously grooved; leaves basal and cauline, membranaceous, green on both surfaces, the basal reniform to ovate, acute to obtuse to the apex, rounded or somewhat tapering to the base, up to 50 cm long, nearly as wide, with mucronate teeth, glabrous, palmately nerved, long-petiolate, the cauline leaves alternate, ovate, usually acute at the apex, more or less tapering to the base, entire to serrate to dentate, glabrous, short-petiolate or sessile; heads discoid, numerous, borne in corymbs, not subtended by bractlets; involucre cylindrical, up to 12 mm high; phyllaries 5, in 1 series, linear-oblong, obtuse to acute at the apex, with scarious margins, the midvein unwinged; receptacle flat, epaleate; ray flowers absent; disc flowers 5, tubular, white, bisexual, fertile, the corolla 5-lobed, 8–10 mm long; cypselae fusiform, with 4–5 strong ribs, dark brown or purple, 4–5 mm long; pappus of numerous white, scabrous bristles 6–8 mm long.

Common Name: Giant Indian plantain.
Habitat: Woods.
Range: Pennsylvania to Minnesota, south to Oklahoma, Mississippi, and Georgia.
Illinois Distribution: Scattered in most of Illinois.

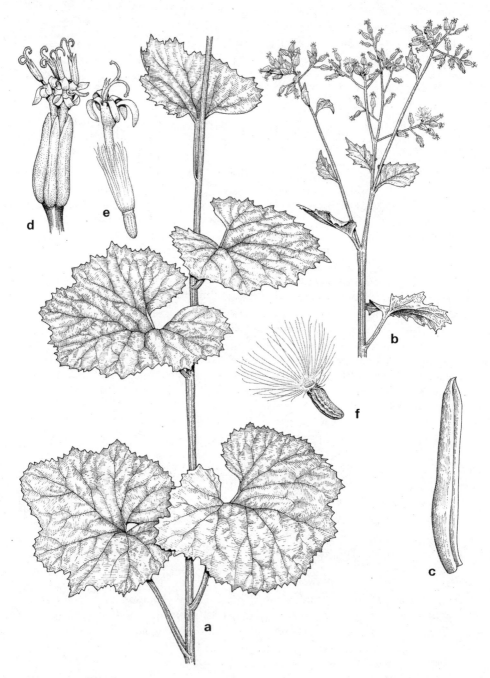

126. *Arnoglossum reniforme* (Prairie Indian plantain).

a. Middle part of plant with leaves.
b. Upper part of plant.
c. Phyllary.
d, e. Disc flowers.
f. Cypsela.

This species is similar in appearance to *A. atriplicifolium* except that the leaves are green on the lower surface rather than glaucous. It usually grows taller than the other species of *Arnoglossum* in Illinois.

This species has a complicated nomenclature, as evidenced by the number of names given to it. For many years, this plant was called *Cacalia muehlenbergii*. That binomial was based on Schultz-Bipontius's *Senecio muehlenbergii*, which was superfluous when published.

Arnoglossum reniforme flowers from July to September.

3. **Arnoglossum plantagineum** Raf. Fl. Ludov. 65. 1817. Fig. 127.
Cacalia tuberosa Nutt. Gen. 2:138. 1818.
Mesadenia plantaginea (Raf.) Raf. New Fl. 4:79. 1836.
Mesadenia tuberosa (Nutt.) Britt. in Britt. & Brown, Ill. Fl. N.E. U.S. 3:474. 1898.
Cacalia plantaginea (Raf.) Shinners, Field & Lab. 18:81. 1950.

Perennial herbs with a thick tuberous root; stems erect, unbranched, to 1 m tall, usually angular, grooved, glabrous; leaves firm, basal and cauline, the basal lance-ovate to oval to broadly elliptic, acute to obtuse at the apex, tapering to the petiolate base, usually entire, glabrous, with parallel venation, green on both surfaces, to 15 cm long, to 10 cm wide, the cauline leaves elliptic, acute at the apex, more or less rounded at the base, entire or serrulate, glabrous, the uppermost sessile; heads discoid, numerous, borne in corymbs, not subtended by bractlets; involucre cylindrical, 10–13 mm high; phyllaries 5, in 1 series, equal, ovate to oblong, with scarious margins, the midvein winged; receptacle flat, epaleate; ray flowers absent; disc flowers 5, tubular, white, bisexual, fertile, the corolla 5-lobed, up to 10 mm long; cypselae fusiform, dark brown, glabrous, several-ribbed, 4–5 mm long; pappus of white, scabrous bristles to 8 (–9) mm long.

Common Name: Prairie Indian plantain.
Habitat: Prairies, springy ground in marshes and bogs, prairie fens, gravelly hill prairies.
Range: Ontario to Minnesota to South Dakota, south to Texas and Alabama.
Illinois Distribution: Widely scattered in Illinois, but not common.

The leaves of this species are so firm that they almost feel like cardboard. The leaves, which have five to nine parallel veins, may be entire or serrulate. This is the only species of *Arnoglossum* in Illinois that has a thickened, tuberous root and phyllaries with a winged midvein.

The specific epithet *plantagineum* predates *tuberosum* by one year.

Arnoglossum plantagineum flowers from June to August.

70. **Tussilago** L.—Coltsfoot

Monoecious perennial herbs with rhizomes, forming colonies; stems erect, unbranched; leaves basal and cauline, alternate, palmately nerved; head radiate, solitary, with indistinct bractlets; involucre cylindrical to campanulate; phyllaries

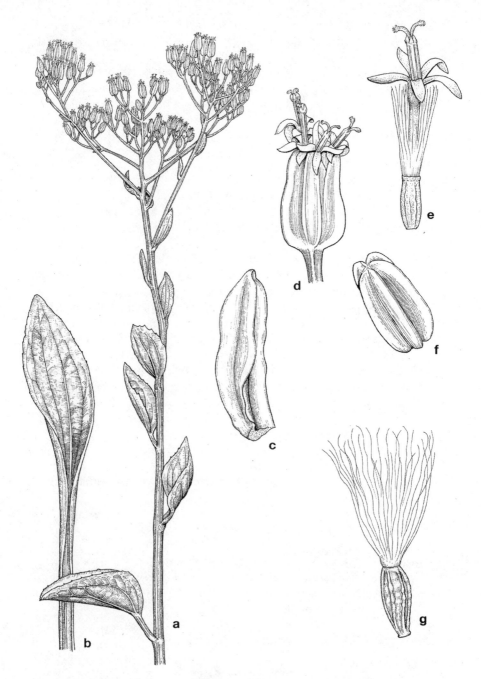

127. *Arnoglossum*
    plantagineum (Giant
    Indian plantain).

a. Upper part of plant.
b. Leaf.
c. Phyllary.
d. Involucre of flowers.

e. Disc flower.
f. Cypsela.
g. Cypsela with pappus.

about 21, usually in 1 series, more or less equal, persistent, with scarious margins; receptacle more or less flat, epaleate; ray flowers numerous, yellow, pistillate, fertile; disc flowers up to 40, tubular, yellow, staminate, the corolla 5-lobed; cypselae cylindrical to prismatic, prominently ribbed; pappus of up to 100 white, barbellulate, caducous bristles.

Tussilago consists of a single species native to Europe, Asia, and north Africa. It is similar to *Petasites*, but differs by being monoecious and by having a solitary head of yellow flowers.

1. **Tussilago farfara** L. Sp. Pl. 2:865. 1753. Fig. 128.

Perennial herbs with rhizomes, forming colonies; stems erect, unbranched, to 50 cm tall, tomentose, coming up in early spring; leaves basal and cauline, the basal appearing after the flowering head, orbicular to broadly ovate, subacute at the apex, subcordate at the base, white-tomentose on the lower surface, at least when young, green above, dentate or shallowly lobed, palmately nerved, to 25 cm long, nearly as wide, on long petioles; cauline leaves alternate, linear to narrowly ovate, bract-like, to 2.5 cm long, tomentose; head radiate, solitary, to 2 cm across, with up to 15 rather indistinguishable bractlets; involucre more or less cylindric; phyllaries about 21, in 1 series, more or less equal, 7–15 mm long, linear, the margins scarious; receptacle flat, epaleate; ray flowers numerous, yellow, to 10 mm long, pistillate, fertile; disc flowers up to 40, tubular, yellowish, staminate, the corolla 5-lobed, 10–12 mm long; cypselae angular or prismatic, 5- or 10-ribbed, glabrous, 3–4 mm long; pappus of numerous white, caducous, barbellulate bristles 8–12 mm long.

Common Name: Coltsfoot.
Habitat: Wooded ravine (in Illinois).
Range: Native to Europe, Asia, and north Africa; adventive mostly in the northeastern and midwestern United States.
Illinois Distribution: Known only from Lake County where it was first collected in 1982.

This adventive species differs from *Petasites hybridus* by its solitary yellow flowering head.

Tussilago farfara flowers during April and May.

71. **Petasites** Mill.—Sweet Coltsfoot; Butterbur

Polygamodioecious perennial herbs with rhizomes; stems erect, unbranched, those with staminate heads withering after flowering, those with pistillate heads elongating after flowering; leaves basal and cauline, the basal petiolate, the cauline alternate and sometimes bract-like; heads discoid (in Illinois) or radiate, several in racemes or panicles, subtended by up to 5 bractlets, some flowering heads staminate, some pistillate and fertile; involucre obconic or turbinate; phyllaries up to 15, usually in 2 series, subequal, with scarious margins; ray flowers absent (in Illinois) or numerous; disc flowers up to 175, tubular, pinkish or purplish (in

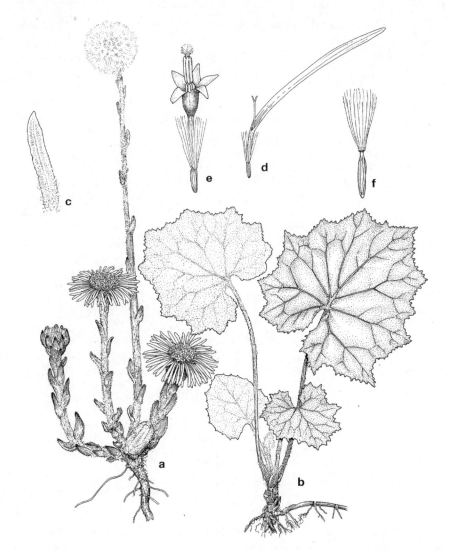

128. *Tussilago farfara*
(Coltsfoot).

a. Habit.
b. Basal leaves.
c. Phyllary.

d. Ray flower.
e. Disc flower.
f. Cypsela.

Illinois), pistillate, fertile, the corolla 5-lobed; cypselae cylindrical to prismatic, with 5 or 10 ribs; pappus of up to 100 smooth or barbellulate white bristles.

There may be as many as eighteen species in the genus, most of them native to Europe and Asia. One adventive species occurs in Illinois.

1. **Petasites hybridus** (L.) Gaertn., Mey., & Scherb. Oekon. Fl. Wetterau 3:184. 1802. Fig. 129.
Tussilago hybridus L. Sp. Pl. 2:866. 1753.

Perennial herbs with rhizomes, sometimes forming colonies; stems erect, unbranched, to 75 cm tall, glabrous or nearly so; leaves basal and cauline, the basal reniform to orbicular, obtuse at the apex, cordate at the base, shallowly lobed, unevenly serrate, glabrous, to 50 cm long, nearly as wide, on long petioles, the cauline alternate, smaller and bract-like, usually unevenly serrate; heads discoid, several in a compact raceme or panicle, subtended by up to 5 bractlets; involucre usually obconic; phyllaries up to 15, usually in 2 series, linear, subequal, 6–15 mm long, with scarious margins; receptacle flat, epaleate; ray flowers absent; disc flowers numerous, tubular, pinkish or purplish, bisexual, fertile, the corolla 5-lobed; cypselae prismatic, with 5 or 10 ribs, glabrous, 3–4 mm long; pappus of numerous white, smooth or barbellulate bristles 8–12 mm long.

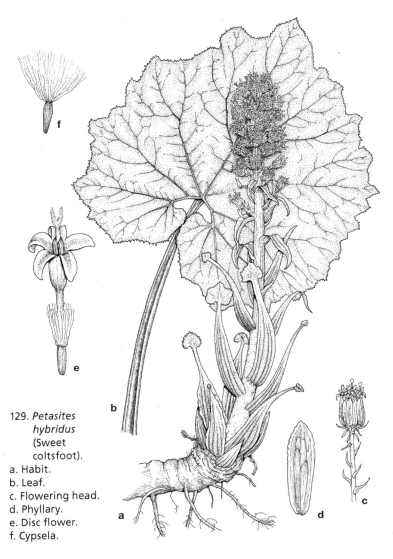

129. *Petasites hybridus* (Sweet coltsfoot).
a. Habit.
b. Leaf.
c. Flowering head.
d. Phyllary.
e. Disc flower.
f. Cypsela.

Common Name: Sweet coltsfoot; butterbur.

Habitat: Ravine (in Illinois).

Range: Native to Europe; escaped from cultivation but rarely established in the United States.

Illinois Distribution: Known from Lake County where a huge colony occurs, first collected in 1968 by Elizabeth Lunn. It has also escaped from cultivation at the Morton Arboretum in DuPage County.

This garden ornamental differs from *Tussilago farfara* by its several flowering heads that are pinkish or purplish and its nearly glabrous stems.

Petasites hybridus flowers during April and May.

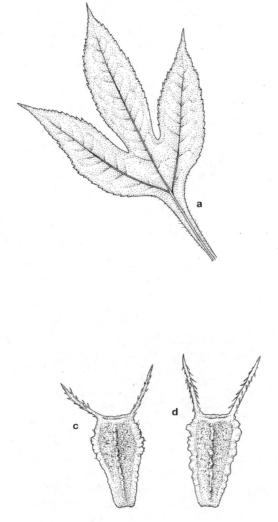

130. *Ambrosia trifida*
    var. *trifida* (Giant
    Ragweed).
a. Leaf. *Bidens polylepis* f.
    retrorsa (Many-bracted
    bur marigold).
b. Cypsela. *Bidens aristosa*
    f. *aristosa* (Swamp
    marigold).
c. Cypsela. *Bidens aristosa*
    f. *fritcheyi* (Swamp
    marigold).
d. Cypsela.

Excluded Species

Actinea acaulis (Pursh) Spreng. Cronquist in 1952 called *Tetraneuris herbacea* Greene by this binomial.

Actinella acaulis (Pursh) Nutt. Clute in 1904 called *Tetraneuris herbacea* Greene by this binomial.

Bidens chrysanthemoides Michx. Many early Illinois botanists used this binomial for *Bidens cernua* L., which is a different species.

Coreopsis auriculata L. Several early Illinois botanists used this binomial for *Coreopsis pubescens* L.

Coreopsis senifolia Michx. Short (1845) and Lapham (1857) erroneously used this binomial for *Coreopsis palmata* Nutt.

Helianthus ambiguus (Torr. & Gray) Britt. Watson in 1929 used this binomial for *Helianthus grosseserratus* M. Martens.

Helianthus atrorubens L. Cronquist in 1952 used this binomial for *Helianthus silphioides* Nutt.

Helianthus parviflorus HBK. Gray in 1884 used this binomial for *Helianthus microcephalus* Torr. & Gray.

Helianthus tracheliifolius Mill. Many early Illinois botanists used this binomial for *Helianthus strumosus* L.

Senecio smallii Britt. This southeastern species was attributed to Illinois by Pennell in 1931, but I have not seen a specimen to verify this.

Silphium connatum L. Michaux in 1803 used this binomial for *S. integrifolium* L.

Xanthium canadense Mill. Several botanists in Illinois used this binomial for *Xanthium chinense* Mill. These are two different species.

Xanthium echinatum Murray. Lapham (1857), Deam (1910), and Pepoon (1927) used this binomial for *Xanthium commune.* These are two different species.

Glossary

acuminate. Gradually tapering to a long point.

acute. Sharply tapering to a point.

annual. A plant that lives for only one growing season.

anthesis. Flowering time.

apiculate. Abruptly short-pointed at the tip.

appressed. Lying flat against the surface.

aristate. Bearing an awn.

array. An arrangement of flowering heads in an inflorescence in the Asteraceae.

attenuate. Gradually becoming narrowed.

auriculate. Bearing an ear-like process.

awn. A bristle usually terminating a structure.

axil. The junction of two structures.

axillary. Borne in an axil.

barbellate. Bearing barbs or setae with downward-pointing prickles.

beaked. Having a short point at the tip.

biennial. A plant that completes its life cycle in two years and then perishes.

bisexual. Said of a flower bearing both staminate and pistillate parts.

bract. An accessory structure on the peduncle bearing flowering heads.

bracteate. Bearing one or more bracts.

bracteole. A secondary bract.

bracteolate. Bearing one or more bracteoles.

bractlet. A small bract.

bristle. A stiff hair or hair-like growth; a seta.

caducous. Falling away early.

campanulate. Bell-shaped.

canescent. Grayish-hairy.

capillary. Thread-like.

capitate. Forming a head.

capitulum. A flowering head in the Asteraceae.

caudex. The woody base of a perennial plant.

cauline. Belonging to a stem.

cespitose. Growing in tufts.

chaff. A scale or group of scales.

chartaceous. Papery in texture.

cilia. Marginal hairs.

ciliate. Bearing cilia.

ciliolate. Bearing small cilia.

clasping. Said of a leaf whose base wraps partway around the stem.

columnar. Shaped like a column or cylindrical upright structure.

compressed. Flattened.

concave. Curved on the inner surface; opposite of convex.

convex. Curved on the outer surface; opposite of concave.

cordate. Heart-shaped.

coriaceous. Leathery.

corolla. That part of a flower composed of petals.

corymb. A type of inflorescence where the pedicellate flowers are arranged along an elongated axis but with the flowers all attaining about the same height.

corymbiform. Shaped like a corymb.

crenate. With round teeth.

cuspidate. Terminating in a very short point.

cyme. A type of broad and flattened inflorescence in which the central flowers bloom first.

cymose. Bearing a cyme.

cypsela. The fruit of a member of the Asteraceae family.

deciduous. Falling away.

decumbent. Lying flat but with the tip ascending.

decurrent. Adnate to the petiole or stem and then extending beyond the point of attachment.

dentate. With sharp teeth, the tips of which project outward.

denticulate. With small, sharp teeth, the tips of which project outward.

diffuse. Loosely spreading.

disc. The group of flowers that makes up the center of a flowering head in some Asteraceae.

disciform. Flowering heads with only disc flowers but with peripheral flowers that have filiform corollas that are usually pistillate.

discoid. Bearing only disc flowers.

eciliate. Without cilia.

eglandular. Without glands.

ellipsoid. Referring to a solid object that is broadest at the middle, gradually tapering to both ends.

elliptic. Broadest at the middle, gradually tapering to both ends.

entire. Without any projections along the edge.

epaleate. Without palea or outgrowths from the receptacle in the Asteraceae.

epunctate. Without dots.

erose. With an irregularly notched tip.

fascicle. Cluster.

fibrous. Referring to roots borne in tufts.

filiform. Thread-like.

flexuous. Zigzag.

fusiform. Spindle-shaped.

glabrate. Becoming smooth.

glabrous. Without pubescence or hairs.

gland. An enlarged, usually spherical body functioning as a secretory organ.

glandular. Bearing glands.

glaucous. With a whitish covering that may be rubbed off.

globose. Round; globular.

glomerule. A small, compact cluster.

glutinous. Covered with a sticky secretion.

hirsute. Bearing stiff hairs.

hirtellous. Finely hirsute.

hispid. Bearing rigid hairs.

hyaline. Transparent.

inferior. Referring to the position of the ovary when it is surrounded by the adnate portion of the floral tube or is embedded in the receptacle.

inflorescence. A cluster of flowers.

involucre. The collection of phyllaries around a flower cluster in the Asteraceae.

involute. Rolled inward.

keel. A ridge-like process.

laciniate. Divided into narrow, pointed divisions.

lamina. A blade.

lanceolate. Lance-shaped; broadest near the base, gradually tapering to the narrower apex.

lanceoloid. Referring to a solid object that is broadest near the base, gradually tapering to the narrower apex.

latex. Milky sap.

leaflet. An individual unit of a compound leaf.

ligule. A flat, narrow, petal-like structure of a flower in the Asteraceae family; a ray.

ligulate. Flowers with ligules.

linear. Elongated and narrow in width throughout.

lustrous. Shiny.

mucro. A short, abrupt tip.

mucronate. Possessing a short, abrupt tip.

mucronulate. Possessing a very short, abrupt tip.

obconic. Reverse cone-shaped.

oblanceolate. Reverse lance-shaped; broadest at apex, gradually tapering to narrow base.

oblong. Broadest at the middle and tapering to both ends, but broader than elliptical.

oblongoid. Referring to a solid object that is broadest at the middle and tapering to both ends.

obovate. Broadly rounded at the apex, becoming narrowed below.

obovoid. Referring to a solid object that is broadly rounded at the apex, becoming narrowed below.

obtuse. Rounded at the apex.

orbicular. Round.

oval. Broadly elliptical.

ovate. Broadly rounded at base, becoming narrowed above; broader than lanceolate.

ovoid. Referring to a solid object that is broadly rounded at base, becoming narrowed above.

palea. An outgrowth from the receptacle in the Asteraceae.

paleate. Bearing paleae.

palmate. Divided radiately, like the fingers from a hand.

panicle. A type of inflorescence composed of several racemes.

paniculate. Having the flowers in a panicle.

pannose. Having the texture of felt.

pappus. Various kinds of structures attached to the cypsela in the Asteraceae.

pedicel. The stalk of a flower.

pedicellate. Said of a flower that has a pedicel.

peduncle. The stalk of an inflorescence.

perennial. Living more than two years.

perfect. Said of a flower that has both stamens and pistils.

petiolate. Having a petiole.

petiole. The stalk of a leaf.

phyllary. A bract subtending a flowering head in the Asteraceae.

pilose. Bearing soft hairs.

pinna. A primary division of a compound blade.

pinnate. Divided once into distinct segments.

pinnatifid. Said of a simple leaf or leaf-part that is cleft or lobed only partway to its axis.

pinnatisect. Divided in a pinnate manner.

pistillate. Bearing pistils but not stamens.

plumose. Bearing fine hairs, like the plume of a feather.

prostrate. Lying flat.

puberulent. Bearing minute hairs.

pubescent. Bearing some kind of hairs.

punctate. Dotted.

pustulate. Having small, pimple-like swellings.

pyramidal. Shaped like a pyramid.

raceme. A type of inflorescence where pedicellate flowers are arranged along an elongated axis.

radiate. Bearing ray flowers in the Asteraceae.

ray. A flat flower in the Asteraceae; a ligule.

receptacle. A structure to which all flowers are attached in the capitulum in the Asteraceae.

reflexed. Turned downward.

resinous. Producing a sticky secretion or resin.

reticulate. Resembling a network.

rhizome. An underground horizontal stem bearing nodes, buds, and roots.

rhizomatous. Bearing rhizomes.

rosette. A cluster of leaves in a circular arrangement at the base of a plant.

rugose. Wrinkled.

rugulose. With small wrinkles.

sagittate. Shaped like an arrowhead.

scabrellous. Slightly rough to the touch.

scabrous. Rough to the touch.

scarious. Thin and membranous.

secund. Borne on one side.

septate. With dividing walls.

sericeous. Silky; bearing soft, appressed hairs.

serrate. With teeth that project forward.

serrulate. With very small teeth that project forward.

sessile. Without a stalk.

seta. Bristle.

setaceous. Bearing bristles or setae.

setose. Bearing setae.

spatulate. Oblong but with the basal end elongated.

spicate. Bearing a spike.

spike. A type of inflorescence where sessile flowers are arranged along an elongated axis.

spinescent. Becoming spiny.

spinose. Bearing spines.

spinulose. Bearing small spines.

squarrose. With tips turned back or spreading.

staminate. Bearing stamens but not pistils.

stipe. A stalk.

stipitate. Bearing a stalk.

stolon. A slender, horizontal stem on the surface of the ground.

stoloniferous. Bearing stolons.

stramineous. Straw-colored.

striate. Marked with grooves.

strigillose. Bearing short, appressed, straight hairs.

strigose. Bearing appressed, straight hairs.

subacute. Nearly tapering to a short point.

subcordate. Nearly cordate.

suborbicular. Nearly round.

subulate. With a very short, narrow point.

succulent. Fleshy.

terete. Round in cross-section.

ternate. Divided three times.

thyrse. A mixed inflorescence containing panicles and cymes.

thyrsoid. Bearing a thyrse.

trigonous. Triangular in cross-section.

truncate. Abruptly cut across.

turbinate. Top-shaped; shaped like a turban.

undulate. Wavy.

urceolate. Urn-shaped.

villous. Bearing long, soft, slender, unmatted hairs.

viscid. Sticky.

whorl. An arrangement of three or more structures at a point.

Literature Cited

Clute, W. N. 1904. *Actinella acaulis. American Botanist* 6:96.

Cronquist, A. 1952. In H. A. Gleason, *Illustrated Flora of the Northeastern United States*, vol. 3. New York Botanical Garden. Bronx, New York.

Deam, C. C. 1912. Additions to the flora of the lower Wabash Valley. *Proceedings of the Indiana Academy of Science* 1911:365–69.

Gray, A. 1884. *Synoptical Flora of North America: Gamopetalae after Compositae* 2:1–402.

Lapham, I. A. 1957. Catalogue of the plants of the state of Illinois. *Transactions of the Illinois State Agricultural Society* 2:492–550.

Michaux, A. 1803. *Flora Boreali Americana.* Published by the author. Paris, France.

Pennell, F. W. 1931. On some critical species of the serpentine barrens. *Bartonia* 12:1–23.

Pepoon, H. S. 1927. An annotated flora of the Chicago region. *Bulletin of the Chicago Academy of Science* 8:1–554.

Short, C. W. 1845. Observations on the botany of Illinois. *Western Journal of Medicine and Surgery* 3:185–98.

Watson, E. E. 1929. Contribution to a monograph of the genus Helianthus. *Papers of the Michigan Academy of Science* 9:305–475.

Wilhelm, G., and L. Rericha. 2017. *Flora of the Chicago Region.* Indiana Academy of Science.

Index of Scientific Names

Names in roman type are accepted names, while those in italics are synonyms and are not considered valid.

Index to Common Names

Robert H. Mohlenbrock taught botany at Southern Illinois University Carbondale for thirty-four years. Since his retirement, he has served as senior scientist for Biotic Consultants, teaching wetland identification classes around the country. Among his more than seventy books are *Vascular Flora of Illinois* and *Field Guide to U.S. National Forests*.